Praise for

The Market Gardener

This is a thorough farming manual that lays out a human-scale farming system centered on good growing practices and appropriate technology. Had I read this book when I was a starting farmer, I would now be farming with a walking tractor on an acre and hailing Jean-Martin as my market gardening guru! This book is going to inspire new farmers to stay small and farm profitably.

— Dan Brisebois, author,
Crop Planning for Organic Vegetable Growers,
farmer at Tournesol cooperative farm, Canada.

Jean-Martin's book is very well done and should be of great use to market growers everywhere. Exchange of ideas and information is so important because when we pass ideas on, the next person gets to start where we got to and take the ideas to another level.

— Eliot Coleman, organic farming pioneer and author,
The Winter Harvest Handbook

The Market Gardener is a very technical yet practical book. What Jean-Martin has done with his micro-farm requires a great deal of planning, good management practices and tough full reflections about new (renewed) horticultural practices, which he shares generously. For ether the home or market garden this book might be as useful as… la grelinette!

— Joseph Templier, French master grower and co-author,
ADABIO guide de l'auto-construction

In France, *The Market Gardener* has quickly become a book of reference for small-scale farming. Both visionary and practical, it is a work of rare intelligence. By sharing a way to work the land for abundance of growth in respect of ecological principles, Jean-Martin offers a new way of connecting to the earth and we thank him for it.

— Charles Herve-Gruyer,
Permaculture teacher and grower at
la Fermedu BecHellouin, France

How do we encourage a new generation of ecological, small-scale farmers? By showing that farming can be a viable, stimulating, and respected career choice. This book offers the hope that a small-scale diversified market garden can be both profitable and personally fulfilling and then goes on to give practical advice on just how to do it. I would offer this book to any new or wannabe vegetable farmer as well as to my seasoned mentors. I can't wait to see how the practices I've read about manifest on my own farm this coming growing season and in the years to come. This is an important new book in my farm library.

— Shannon Jones,
small-scale organic market gardener
Broadfork Farm, River Hebert, NS

This is a fantastic addition to any aspiring market gardener's library, and even has a few new ideas for old hands. Jean-Martin has laid out all of the basics for how we can farm more profitability, productively, and passionately on a more human sized scale. This book goes beyond the theoretical, providing valuable details from his own market garden and his experiences over the years. All of this is made even more valuable because of his acknowledgement of the importance of place and also that there is an evolution to any endeavor. Grounding us with an explanation of his own small farm history and location makes it easier for us to learn from his experiences and apply them to our own small farms.

— Josh Volk,
Slow Hand Farm, Portland, Oregon

Jean-Martin Fortier extols the virtues of being small-scale, and expertly details the use of such scale-appropriate tools as broadforks, seeders, hoes, flame weeders, low tunnels, high tunnels, and many other unique tools, specifically designed for this brand of farming. He picks up right where Eliot Coleman has left us, applying many of his core principles, but doing it in such a brilliant way as to provide beginning farmers a solid framework of the information they need to start up and become successful small-scale organic growers themselves.

— Adam Lemieux,
Product Manager of Tools & Supplies
Johnny's Selected Seeds

Jean-Martin Fortier takes our hands and our hearts in his, as he recounts the lessons, practices and motivations behind his incredibly productive and profitable market garden business.

As he leads us through his packing, potting and work sheds, his greenhouses, his fields and his markets we come to know the grounded reasons behind his choices, the surprisingly relaxed rhythm of their lifestyle and work and the simple yet efficient techniques he and his partner employ on their farm. Through his tale, he inspires new and older famers alike to continue to learn how to farm better, and to continue to question the logic of getting 'bigger'.

In his frank, unassuming style, Jean-Martin creates an infallible argument for the sound economics and the appealing lifestyle of his small farm operation. And as he shares all of his farms "secrets of success" he convinces us that anyone—who is smart, determined and hard-working—can build a farm like his.

As Jean-Martin points out, new farmers today have both the choice and ability to build viable small farm operations. But, as he places their choices in the context of a world with increasing complex and fragile ecological, food and financial systems, with the distance between each other and our natural world growing, it is clear that farmers not only have the choice but they have an imperative to take up the calling and build meaningful farm livelihoods that will continue to sustain themselves and all of us.

— Christie Young,
Founder and Executive Director of FarmStart

the market gardener

A SUCCESSFUL GROWER'S HANDBOOK
for SMALL-SCALE ORGANIC FARMING

Jean-Martin Fortier

Foreword by Severine von Tscharner Fleming, The Greenhorns
Illustrations by Marie Bilodeau

new society
PUBLISHERS

Copyright © 2014 by Jean-Martin Fortier. All rights reserved.

© Les Éditions Écosociété, 2012, for the original French edition, *Le jardinier-maraîcher,*
Manuel d'agriculture biologique sur petite surface. www.ecosociete.org

Cover design by Diane McIntosh. Illustration by Marie Bilodeau. All interior illustrations by Marie Bilodeau.

English translation by Scott Irving, Edgar Translation. www.edgar.ca

Printed in Canada. Fourth printing January 2015.

New Society Publishers acknowledges the financial support of the Government of Canada
through the Canada Book Fund (CBF) for our publishing activities.

Inquiries regarding requests to reprint all or part of *The Market Gardener*
should be addressed to New Society Publishers at the address below.

To order directly from the publisher, please call toll-free
(North America) 1-800-567-6772, or order online at www.newsociety.com

Any other inquiries can be directed by mail to:
New Society Publishers, P.O. Box 189, Gabriola Island, BC V0R 1X0, Canada, (250) 247-9737

LIBRARY AND ARCHIVES CANADA CATALOGUING IN PUBLICATION

Fortier, Jean-Martin, 1978–
[Jardinier-maraîcher. English]
The market gardener : a successful grower's handbook for small-scale
organic farming / Jean-Martin Fortier ; foreword by Severine von Tscharner
Fleming, The Greenhorns ; illustrations by Marie Bilodeau.

Translation of: Le jardinier-maraîcher.
Includes bibliographical references and index.
Issued in print and electronic formats.
ISBN 978-0-86571-765-7 (pbk.).—ISBN 978-1-55092-555-5 (ebook)

1. Truck farming. 2. Organic farming. 3. Permaculture. I. Title.
II. Title: Jardinier-maraîcher. English.

SB321.F6913 2014   635'.0484   C2013-907877-0
                                             C2013-907878-9

New Society Publishers' mission is to publish books that contribute in fundamental ways to building an ecologically sustainable and just society, and to do so with the least possible impact on the environment, in a manner that models this vision. We are committed to doing this not just through education, but through action. The interior pages of our bound books are printed on Forest Stewardship Council®-registered acid-free paper that is **100% post-consumer recycled** (100% old growth forest-free), processed chlorine-free, and printed with vegetable-based, low-VOC inks, with covers produced using FSC®-registered stock. New Society also works to reduce its carbon footprint, and purchases carbon offsets based on an annual audit to ensure a carbon neutral footprint. For further information, or to browse our full list of books and purchase securely, visit our website at: www.newsociety.com

Contents

DEDICATION

Dedicated to the organic pioneers
who have paved the way before us. With deep gratitude.

And to young agrarians who are now changing the face of agriculture.
We have not only the choice to do things differently, but the means as well.

Acknowledgments to the English edition of Le Jardinier-Maraîcher

WRITING THIS BOOK WAS NO SMALL FEAT, and I have sunk countless hours into it. This writing adventure would never have been possible without the support of my family, the collaboration of our farm employees, and all the many volunteers who passed through to lend a hand. The long Quebec winter also played its part…

Different people have given their time to help in reviewing this book. I would especially like to thank my long-time friend Kory Goldberg for his dedicated and reflective comments and for his help in editing the manuscript. Ian LeCheminant's sharp copy-editing, Scott Irving's skill in translation, and John McKercher's proficiency in layout and design have all contributed to this final product. Working with such talented people is a real blessing.

This book would not have been possible without the help of FarmStart Ontario, an organization dedicated to supporting a new generation of entrepreneurial, ecological farmers. Their crowdfunding project kick-started this translation. Special thanks to Christie Young for taking on the idea of bringing my work to a broader audience.

I offer my thanks to Severine von Tscharner Fleming, who so kindly agreed to write the foreword. It is the work of people like her that fosters a hospitable climate for us small-scale growers.

Finally, I would like to renew my recognition to everyone who helped out with the original French edition. Naming everyone here would be too long, but I cannot express enough gratitude to Marie Bilodeau for her great art work, Laure Waridel for such an inspiring foreword, and the whole team at Écosociété who believed in this book from the very beginning. The success of *Le Jardinier-Maraîcher* is a result of your collaboration. Merci.

In closing, I would like to thank two people for contributing to the person I am. First, my father, who taught me at a very young age the importance of being well-organized. This has been the best arrow in my quiver. And finally, thanks to Maude-Hélène Desroches, my work partner, my best friend, and the love of my life.

— Jean-Martin Fortier

FarmStart is a charitable organization in Canada that provides tools, resources, and support to help a new generation of entrepreneurial, ecological farmers to get their farms off the ground and to thrive. We need young farmers, new farmers, and more farmers to revitalize our rural communities, root resilient and sustainable food systems, and provide careful stewardship of our agricultural resources for generations to come.

We can't make starting a farm easy, but Farm-Start works to make it a little less risky, a little more accessible, and a lot less lonely.

In the winter of 2013, FarmStart began an online fundraising campaign to translate *Le Jardiner-Maraîcher*, the original French version of *The Market Gardener*. We felt it was important to make available in English this transformational guidebook that can inspire young and new vegetable farmers. This book provides practical information not only on how to start a market garden enterprise, at an affordable and accessible scale, but most importantly about how to make it both very productive and profitable.

We are thrilled that now a new generation of farmers will be able to read and reread what will be a beloved must-have in a vegetable grower's library. We are excited to see how farmers will adopt and continue to adapt the ideas, techniques, and practices that Jean-Martin and Maude Hélène have proven to work on their farm and have shared with us in this fantastic guidebook.

We are grateful to all the donors who supported the fundraising campaign and made this translation possible. And we are thankful for all the passionate, pioneering, and dedicated farmers who are finding innovative and sustainable ways to grow delicious food on less land and with fewer resources. To have vibrant and resilient food systems in the future, we will need more of them.

Foreword
by Severine von Tscharner Fleming, The Greenhorns

TAKING THE FIRST STEP TOWARDS a better future is always the most difficult. Forty years ago, economist E.F. Schumacher offered us *Small Is Beautiful*, an accessible and appropriate treatise to help us take such a step in the chaotic global economy. Poet and agrarian philosopher Wendell Berry suggested that "there is no big solution," only many small ones, and that we must rebuild the economy from the ground up.

I met Jean-Martin at a Greenhorns Grange hall mixer in the Adirondacks of New York. He arrived with his wife, Maude-Hélène, and their two lively, delightful children in a vegetable delivery van piled high with bicycles and camping gear. After the workshops on oxen, soil life, and fermentation, followed by a puppet show, dance party, and pig roast, the whole family quietly mounted their bikes and returned to the tent they'd set up in a nearby field. There were quite a few words spoken about these charming Canadian interlopers, and we made sure to get them on our mailing list. I later visited the couple's farm in rural Quebec, about 60 minutes north of Burlington, Vermont. Beyond the incredible gardens, I was astounded that the farm had as much recreational gear as farming equipment!

And most of it could probably be stuffed into and onto their big white van. Talk about low-tech.

Les Jardins de la Grelinette is a place of tremendous productivity—the gardens bursting with huge cabbages, humming bees, and wheelbarrows darting in and out of permanent vegetable beds neatly tucked into their remay and black mesh. The couple has transformed a derelict rabbit barn into a compelling, comfortable, beautiful home, farm and workspace. La Grelinette is a place of beauty with its ample wild berries, ferns, and forest interspaced with swimming holes, hand-built cabins for visiting interns, and a wood-powered sauna. Every aspect is modest, functional, well-designed, and well-considered, with happy farmers to boot! It's a living testament to the opinions and operations described in *The Market Gardener*. They have pulled it off, and so can you!

Aspiring young farmers are currently confronted with tremendous structural odds, such as an economy that undervalues food, but where real estate pressure forces land prices up. In this challenging climate that discourages small business startups, big businesses set the terms of trade and benefit from unfair labor practices, while subsidizing production costs and externalizing environmental ones. Industrial agriculture has dominated the landscape for the last 40 years, polluting the water, and skewing the marketplace, while warming the climate for future generations. The only "price" agribusiness will pay is to lobby hard in order to keep their status quo. These business interests might be mega-sized and intimidating, but let us not underestimate the cumulative power of many small initiatives. Like the humble acorn that grows into a mighty oak, we have the power to grow up from underneath.

My experience documenting and interviewing the growing young farming community over the past 7 years connects me with a dense fabric of personal and professional narratives around farm startup. I have heard hundreds, if not thousands, of personal farm startup stories, from romance to breakup. I believe that the information, advice, and content in this book, based on Fortier's experience, is invaluable precisely because it is approachable and doable without a lot of money, or land, or debt, or infrastructure—major stumbling blocks for a young person to confront. A frustrated farm apprentice, evaluating these and other challenges, may decide to drop out of agriculture to pursue a more secure, reliable income in other fields. By laying out a micro-sized enterprise, Jean-Martin is not only giving these aspirants the how-to of vegetable production, but has laid out an accessible, simple economic plan that interprets the feasibility of success in small-scale organic farming. This in turn represents a powerful leverage point in increasing the numbers of farmers overall, as well as pointing to a "way in" of economic opportunity for the rest of us.

The corporate food system is now fully centralized and controls many factors that undermine the sector's ultimate resilience. It is energy dependant, highly concentrated, and ultimately unsustainable on any long-term evaluation. Unfortunately, it also controls much of the land base. Where then, in this landscape of monocultures and degraded soils, are the spaces of opportunity? We have already seen good economic traction from CSAs and farmers markets mushrooming all over the US, Canada, and Europe. In some places, especially progressive cities like New York, San Francisco, and Boulder, these markets may seem close to saturated. But in many more areas, these foods are still not available, and the market is untapped.

As I travel around North America, I keep an eagle eye out for the places of strategic opportunity for further farm development. Here, again, Jean-Martin has identified an opening in peri-urban areas, in and around smaller cities and larger towns, especially where the built environment has contracted from the exit of industry, or the breakdown of previous industrial agricultural sectors (e.g., poultry, tobacco, cut flowers, horticulture, equine). In these contexts, there are many small parcels, and small broken-down farm properties perfectly suited to intensive cultivation by La Grelinette-type market gardeners. Vacant urban lands, fractured farming landscapes that have been split up by development, institutionally owned land, and peri-urban marginal lands may be some of the most affordable options for owner-operators, as a full-time or part-time occupation. This could be a "starter farm" that helps the farmers save up money to move operations further out into the countryside after a few years, or else a way to live a farming life in the city—the best of both worlds! The growing strategies and know-how presented in this book are important in this regard.

The Williamson Act, recently enacted in the State of California, is a legislative victory pointing to the potential for transforming the terms of development. Under this legislation, marginal, blighted, or under-utilized lands within the urban boundary can be rented to commercial farmers. If the landlords make a 5-year lease agreement with farmers, the City will waive their property taxes. This kind of law gives new farmers a bargaining chip to be used in negotiations with landowners, and the possibility of small-scale startup. This is only one example among many other initiatives, particularly in urban gardening, urban greenways/urban land trusts, for food security purposes.

It may well prove that micro, low-cost, low-input, high-diversity and high-productivity systems have a major role to play in rebuilding regional food security. The example set out at La Grelinette is living proof of this security. Jean-Martin and Maude-Hélène's work follows in a tradition of appropriate growing practices established by Alan Chadwick, John Jeavons, Eliot Coleman, and Miguel Altieri in the US, as well as Cuban Farmers, the Basque Farmers union, and others from the Via Campesina peasant movements from all over the world. Small-scale biological farming is making a comeback—the right-sized tool for the job—helping individuals and families wiggle free from a dependency on unfair waged labor in the mainstream economy. Small-scale farming is the right tool, precisely because it's compatible with a set of opportunities in our current economy (in both the developing and developed worlds), and is adapted to the possibility of a new economy that will inevitably emerge as conditions (in both energy and transportation), scale, and control are forced to contract.

The modest scale of such operations may not match contemporary culture's obsession with size and economy of scale, but following a different narrative, one that is more suitable in the long run, is both profitable and possible. The guidance given in this book, especially around limiting mechanical investments and overhead costs, might prove to be a more successful business mandate, as well as a more beautiful way to live.

This handbook is a testimony to such ideas, made manifest in a real place. It's an amazing first read for beginners because it's comprehensive, holistic, and succinct. While the nuance and detail of sustainable agriculture may take a lifetime to master, the lessons, experience, and skill-set shared in this volume are sufficient to get started. The straightforward approach, transparent economic considerations, and clear instructions presented by Jean-Martin should enable anyone willing to commit to a few years of apprenticeship and outdoor handcraft adventure a certitude that they can start their own career in farming. And start now—our world needs more farmers!

Severine von Tscharner Fleming is an organizer, filmmaker and farmer living in the Champlain Valley of New York. She runs the Greenhorns, a 6-year-old nonprofit network for young farmers in the US (thegreenhorns.net). Severine is also co-founder and board member of Farm Hack, an open source platform and workshop for appropriate farm technologies (farmhack.net) and co-founder of National Young Farmers Coalition (youngfarmers.org).

Preface

AFTER FINISHING MY UNIVERSITY STUDIES at the McGill School of Environment in Montreal, my wife Maude-Hélène and I set out on a two-year journey to Mexico and the United States to work on small organic farms. Coming from a suburban background disconnected from nature, this newly discovered rural lifestyle changed the way I saw the world. Spending long hours each day outside not only made me rethink my political and philosophical positions, but it nourished my soul. After spending so many years—indoors—reading about how the modern global economic system is destroying our planet's ecological integrity, it felt great to finally find a direct way to impact the world in a positive manner. The farmers and the farming communities where we stayed were amazing and we felt blessed to have the opportunity to take part in their way of life. I had found practical idealism.

Coming back home to integrate our lessons from abroad, Maude-Hélène and I spent a few years as self-employed market gardeners on rented land. We started a family and eventually felt the need to have our own home. By then we knew we wanted to get established in farming. Once we found our ten-acre site in Saint-Armand, in the Eastern Townships of Quebec, we immediately began to put into practice the things we had

learned about permaculture and intensive cropping systems. Soon enough we built up a very productive market garden on less than two cultivated acres. We named the farm after the *grelinette* ("broadfork" in English), a tool that epitomizes efficient hand labor in ecological gardening.

Maude-Hélène and I began this venture together, and the success of our micro-farm is the result of our collective intelligence and hard work. So, while *The Market Gardener* represents my own opinions and suggestions, I use the pronoun "we" throughout the book when describing the horticultural methods and techniques we used on our farm.

The Market Gardener grew out of a desire to provide aspiring farmers with a tool to help them start their businesses. For a number of years I had worked with Montreal's *Équiterre*, a non-profit organization dedicated to sustainable development, serving as a mentor for beginning farmers. It became clear to me that although persistence, determination and hard work are all key ingredients for successful farming, these qualities on their own are not enough. Careful planning and design, good management practices, and appropriate choices of equipment are all essential components for developing an understanding of the farm as a whole system. And since it has been uncommon in Quebec to grow vegetables on a micro-scale level using hand tools, I felt that our experience contained valuable information to pass along.

To this end, I set out to describe the horticultural practices used in our market garden, chapter by chapter, in as much detail as possible. The learning curve in growing crops commercially is steep, and I have always believed that a seasoned grower is in the best position to impart the know-how required for the tasks at hand. Personal experience has also taught me that having a clear guide on what to do at each stage of the growing season, and a good example to follow, are both essential when you don't have much experience in a given field. I believe this handbook provides valuable guidance.

One of the guiding principles in writing this book was to share only growing methods that we successfully practiced for many seasons on our farm. This assures the reader that the information presented is both accurate and proven. This being said, I have not touched upon many other practices and techniques used by other successful growers and I encourage the reader to explore different cropping systems than the one I describe. There are many great books about organic gardening and farming and I have recommended additional texts in the annotated bibliography.

Finally, it's important to state that the practices described in this book and used on our micro-farm are not set in stone. We read voraciously, visit as many farms as possible, and constantly communicate with other growers. From time to time, our research leads us to discover better tools and more effective growing techniques. Our production system is a constant work in progress, and our methods will undoubtedly be further refined with time. Nonetheless, I am confident that if you plan on starting an organic market garden then you will find the accumulated knowledge we are presenting to be a useful point of departure and reference. I wish you the best along your agricultural journey and look forward to hearing how you give shape to your market garden, in new places and in different ways.

— Jean-Martin Fortier,
Saint-Armand, Quebec, September 2013.

Small Is Profitable

Nearly everywhere we look, the stirrings of a revolution are becoming increasingly clear: people are farming differently; and we see signs of landowner resistance with a focus on local production, concern for the environment, and citizenship.

— Hélène Raymond and Jacques Mathé
Une agriculture qui goûte autrement. Histoires de productions locales, de l'Amérique du Nord à l'Europe, 2011

EVERYWHERE AROUND THE WORLD, people's eyes are being opened to the ravages of industrial agriculture: pesticides, GMOs, cancer, agribusiness. Along with this growing awareness is an increasing consumer demand for healthy, local, organic food. Alternative modes of selling and purchasing food are also gaining ground, visible not only in the mushrooming farmers' markets but also through community-supported agriculture, or community-shared agriculture (CSA) schemes. This system is a direct exchange between producers and consumers. The consumer buys a share in the farm's production at the beginning of the season, thus becoming a partner in the endeavor. In exchange, the farm commits to providing quality produce, usually harvested the day before, or even the same day. In addition to issues of quality, this model of food distribution addresses people's desire to have a relationship with the farmers who grow their food.

These ideas are making headway in Quebec:

Équiterre, which oversees one of the largest networks of organic farmers and citizens in support of ecological farming, has brilliantly complemented the notion of the family doctor with that of the "family farmer." Alternative modes of food distribution now represent a growing niche, and moving out to the country to make a living in agriculture is now a viable option for young (and not-so-young) aspiring farmers.

My wife and I began our farming career in a very small market garden, selling our veggies through a farmers' market and a CSA project. We rented a small piece of land ($\frac{1}{5}$ of an acre) where we set up a summer camp. It didn't take much investment in the way of tools and equipment to get us up and running, and our expenses were low enough that we were able to cover our farming costs, earn enough money to make it through the winter, and even do some travelling. Back then we were content just to be gardening and to be making ends meet.

Eventually, however, there came a time when we felt the need to become more settled; we wanted to build a house of our own and put down roots in a community. Our new beginning meant that our market garden would have to generate enough income to make payments on the land, pay for the construction of our house, and keep the family afloat.

To accomplish this, we could have followed a route similar to that taken by all the other growers we knew: invest in a tractor and move towards a more mechanized growing system. Instead, we opted to stay small-scale and continue relying on hand and light power tools. From the outset, we had always believed that it was possible—and even preferable—to intensify production through gardening techniques. To grow better *instead* of bigger became the basis of our model. With simplicity in mind, we began researching horticultural techniques and tools that could make farming on our one-and-a-half-acre plot a viable reality.

After much research and many discoveries, our journey led us to what is now a productive and profitable micro-farm. Every week, our market garden now produces enough vegetables to feed over 200 families and generates enough income to comfortably support our household. Our low-tech strategy kept our start-up costs to a minimum and our overhead expenses low. The farm became profitable after only a few years of production, and we have never felt the pinch of financial pressure. Just like in the beginning, gardening is still our main focus, and even though there have been a lot of changes around the farm over the years, our

lifestyle has remained the same. We don't work for the farm; the farm works for us.

We decided to brand ourselves specifically as market gardeners (*jardiniers-maraîchers* in French) to emphasize the fact that we work with hand tools. Unlike most contemporary vegetable producers, who grow in vast fields, we work in gardens where our fossil fuel input is relatively low. The features that characterize our operation—high productivity on a small plot of land, intensive methods of production, season extension techniques, and selling directly to public markets—are all modelled after the French tradition of maraîchage, although our practices have also been influenced by our American neighbors. The greatest influence on our work has been the American vegetable grower Eliot Coleman, whom we have visited and met on several occasions. His book *The New Organic Grower* guided us and helped us see that it truly is possible to turn a profit on less than two cultivated acres. Coleman's shared experience and his innovation in techniques for growing vegetables on small plots were a gift to us, and we owe him a great deal.

Of course, most established farmers would probably tell us that farming without a tractor is too much work and that we are too young to appreciate how much easier our lives would be with mechanization. I disagree. The cultivation techniques described in this book actually reduce the amount of work required for field preparation, and planting crops more closely together greatly reduces weed pressure. And though most of our gear and tools are hand-powered, they are quite sophisticated and designed to make tasks more efficient and ergonomic. All in all, apart from harvesting, which accounts for the bulk of our work, our productivity and efficiency are extremely high. The manual labour we do is pleasant, lucrative, and very much in keeping with a healthy lifestyle. More often than not, we enjoy the sound of birdsong as we work, rather than the din of engines.

None of this is to say that I object to all forms of mechanization. Of the most successful farms I have visited, the majority were highly mechanized—Eliot Coleman's being the exception. I would simply put it this way: using a tractor and other machinery for weeding and tilling does not by itself guarantee that farming will be more profitable. When choosing between a non-mechanized approach and machinery such as a two-wheeled tractor, aspiring farmers must always weigh the pros and cons, especially if they are just starting out.

Can You Really Live off 1.5 Acres?

When it comes to commercial vegetable growing, the idea of a profitable micro-farm is sometimes met with scepticism by people in the farming world. It is even possible that some naysayers would try to discourage an aspiring farmer from starting an operation like ours, stating that production simply won't be enough to make ends meet for a family. I encourage aspiring farmers to take this kind of scepticism with a grain of salt. Attitudes are beginning to shift as micro-farming in the United States, Japan, and other countries is

demonstrating the impressive potential of biologically intensive cropping systems geared towards direct selling. Our farm in Quebec, Les Jardins de la Grelinette, is living proof of this. In our first year of production on rented land, our farm brought in $20,000 in sales with less than one quarter of an acre under cultivation. The following year, our sales more than doubled on the same garden size, rising to $55,000. In our third growing season, we invested in new tools and land, settling on our own farm site in Saint-Armand. By increasing our area under cultivation to one and a half acres, we were able to increase our gross sales to $80,000. When our sales broke the $100,000 mark the following year, our micro-farm reached a level of production and financial success that most people in the agriculture industry believed to be impossible. When our sales figures were made public through a farming competition, our business won a prize for its outstanding economic performance.

For the last ten years, my wife and I have had no other income than the one we obtain from our 1½-acre micro-farm. Many other small-scale growers make better than a living wage on small intensively cultivated plots, and there should not be any doubt that it is possible to have a career in market gardening. In fact, one can imagine making a pretty decent livelihood. A well-established, smoothly running market garden with good sales outlets can bring in $60,000 to $100,000 per acre annually in diverse vegetable crops. That's with a profit margin of over 40%—a figure that stacks up favorably against margins in many other agricultural sectors.

Our daily life in the garden is in tune with the passing seasons and in line with how we want to live. Market gardening is hard work, but also rewarding and fun.

Not Just Making a Good Living, but Making a Good Life

The popular myth of family farms persists: we are tied down to the land, we work seven days a week, we never have time off, and we just barely scrape by financially. This image probably has its roots in the real-life struggles experienced by most conventional farmers, who are caught in the stranglehold of modern agriculture. It is true that being a mixed vegetable grower is hard work. Rain or shine, we are up against the vagaries of a highly unpredictable climate. Bumper crops and seasons of plenty are far from guaranteed, and a hefty dose of pluck and commitment is required to make it through—particularly during those first few years, when one is still building infrastructure and a customer base.

Our vocation is nevertheless an exceptional one, defined not by the hours spent at work or the money earned, but by the quality of life it affords. Believe it or not, there is still plenty of free time left over when the work is done. Our season gradually gets started in the month of March and finishes in December. That's nine months of work; three months off. The winter is a treasured time for resting, travelling, and other activities. To anyone who pictures farm life as endless drudgery, I would assert that I feel quite fortunate to live in the countryside and work outdoors. Our work offers us the opportunity to become partners with nature on a daily basis, a reality that not many other professional careers can offer. Unlike employees of big companies living with the constant threat of layoffs, I have job security. That's saying a lot.

After having spent so much time at the computer writing this book, I would also add that the physical demands of market gardening are actually easier on one's health than sitting in front of a computer screen all day. By saying so, I hope to reassure some readers that gardening as a living is not so much a question of age as one of will. Whether or not you have a background in farming, you can learn everything you need to know in this time-honored vocation if you are serious and motivated. You need only invest your time and enthusiasm.

Since our farm began hosting interns just getting their feet wet in the world of agriculture, I have noticed that most aspiring farmers I meet are drawn to the fields for one fundamental reason. It's not just that they want to be their own boss and get out in the fresh air as much as possible—most of them are looking for work that brings meaning to their lives. I can understand this, because I have found much fulfillment in being a family farmer. Our toil in the garden is rewarded by all the families who eat our vegetables and thank us personally every week. For anyone looking for a different way of living, market gardening offers a chance not only to make a good living, but also to make a good life.

Succeeding as a Small-Scale Organic Vegetable Grower

> *To obtain the best yield from the soil, without excessive expenses, through the judicious selection of crops, and through appropriate work: such is the goal of the market gardener.*
>
> – J. G. Moreau and J. J. Daverne,
> *Manuel pratique de la culture maraîchère de Paris*, 1845

BECAUSE OUR MICRO-FARM has garnered so much media attention in recent years, farmers of all stripes and many agronomists have been coming to meet us and visit our gardens. These people, most of them only familiar with modern large-scale conventional farming, are curious about our work because we challenge the belief that the small family farm cannot stay afloat in today's economy. Despite our decade of experience in proving the viability of a micro-farm, most of these visitors remain unconvinced. They find it difficult to wrap their heads around the fact that we have no plans to make major investments and that we intend to stay small and continue working with hand tools. A bank loan officer who visited us adamantly declared as she left that we were not real business people, and that our farm was not a real farm!

Our farming choices may be easier to under-stand when one stops to consider the obstacles that beginning farmers must face when they are just getting started. For us, the decision to grow vegetables on a small plot of land, while minimizing start-up investments, simply had to do with our financial reality at the time. When we were in our early twenties, our financial resources were limited and we felt strongly about the importance of minimizing our debt load. Ten years later, our strategy of starting a farm without a large capital expense, while still producing high yields of vegetables for direct sales, has proved to be lucrative. Our market garden demonstrates that high profits can be earned *without* high costs.

For beginning farmers, there are a number of advantages to "starting small"—but there is also much to be said for staying small in the years that follow. That being said, whatever the size of the planned operation, it is important to understand

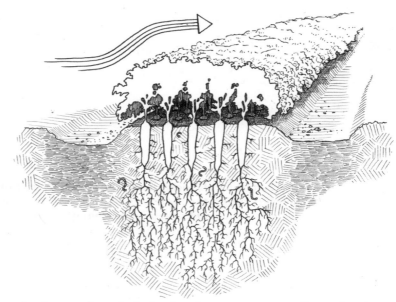

When crops are closely spaced on a bed, the plant leaves come to rapidly touch one another, creating a beneficial microclimate. This canopy reduces weed growth, helps retain moisture in the soil, and protects the crop from wind damage.

In some circles, the word "biointensive" refers to a very narrowly defined set of practices and techniques. Some people have even tried to trademark the approach. I generally prefer the expression "biologically intensive," and I will use it more often in this book, but both refer to the same ideas and principles.

the implications of different production strategies in order to make wise choices about how to best succeed as an organic vegetable grower. This chapter touches on a few factors which, in my opinion, are at the heart of our market gardening success.

A Biologically Intensive Approach

The term "biointensive"* broadly refers to a horticultural method in which growers maximize crop yields from a minimum area of land, while seeking to preserve—or even improve—the quality of the soil. Drawing on the experience of 19th-century French vegetable growers and Rudolph Steiner's biodynamic principles, the biointensive method was refined in northern California beginning in the 1960s.

There is now a whole literature on biologically intensive vegetable growing methods, and although the techniques most often discussed in these works (see bibliography for recommended reading) are geared towards home gardening, a number of the practices can be useful in the context of commercial production. We took one such approach in developing our cropping system.

To begin with, we have not arranged our garden in the rows typically used in mechanized farming where crops are spaced according to the dimensions of the tractors and weeding machinery. Instead, we grow our crops in permanent raised beds. While establishing the beds, we invested in a large quantity of organic matter with the idea of quickly creating a rich and living soil. We effectively *built* our soil this way. Since then,

Mechanized farming, whether conventional or organic, spaces the rows of crops according to the dimensions of the tractors and the weeding machinery. Since we use only hand-powered tools to control weeds on our farm, we do not have this limitation.

we've continued to add compost regularly while limiting any turning of the soil to just the surface, thereby keeping the structure as intact as possible. For deep soil cultivation, we use a broadfork (*grelinette* in French) which allows us to aerate the soil without turning it. The purpose of this cultivation is to create loose, fertile soil, which encourages the crop roots to spread downward rather than sideways. This in turn makes it possible to plant the crops quite close together without them running into each other at the root level.

The goal is to space the crops such that their leaves touch each other when the plants reach three quarters of their full size. At maturity, the foliage will cover all of the growing area, effectively creating a living mulch. This strategy of closely spacing the crops (in addition to allowing

high yields per square foot) has two main advantages. The first is that it greatly cuts down on the amount of weeding required, and the second is that it renders many day-to-day gardening tasks more efficient. These benefits will be explained in detail throughout this book.

In our gardens, it is the quality of the soil structure, combined with the microbe- and nutrient-rich compost that we use, which allows intensive spacing to work well. It took us a few years of trial and error to determine the proper spacing for each crop—so that they are planted as densely as possible without limiting their size at maturity—but it was a worthwhile effort. We also sought to further maximize our growing space by planting as many succession crops as possible. This means that we must determine the length of time each

If the growing area is, for example, five times more densely planted, covering crops with a row cover will take one fifth of the time and use one fifth of the material to do the same job, saving both time and money. Similar efficiencies are also true for irrigation, mulching, and weeding.

crop spends in the garden and plan our seedings so that harvested crops are replaced by new ones as soon as they are out of the garden. Using our crop plan, we succeed in producing multiple successive harvests each season in the same space.

Most of the thinking behind biologically intensive methods is not so very different from the principles of organic agriculture. In both cases, the objective is to build rich, loose, fertile soil. But biointensive practices stress the importance of building soil in order to achieve this. Planting closely spaced crops in permanent beds is what allowed us to establish ourselves in farming without mechanizing our operation. These are not new ideas, and we do not pretend to have invented them. If we can take credit for anything, it's that

we developed a good regime for making our market garden highly productive in a cold Canadian climate while favoring an approach that sustains soil quality.

Minimizing Start-Up Costs

Starting a farm requires investing in tools and equipment, but by starting small and growing crops intensively, it is possible to do so without a large capital outlay. Here is a list of the investments I feel are necessary to run an efficient vegetable operation on less than two acres (1 hectare). The approximate dollar amounts listed are in Canadian funds and are for new equipment which should last many years.

The start-up costs total $39,000. This may sounds like a lot of money to start a micro-farm, but one should consider the following. Firstly, a bank loan of $39,000 spread out over 5 years at 8% interest per year means that the annual investment is about $9,500—which is little enough compared to the potential revenue of a market garden. Of course, these will not be one's only business expenses. This figure does not include certain necessities such as a delivery vehicle, land rental/purchase fees, mortgages, or other variable costs (inputs, administration fees, supplies, etc.). But even so, initial costs are still relatively small, especially when compared to the cost of equipment used in mechanized vegetable growing.

Secondly, some of these items can be purchased secondhand or gradually with time. We were fortunate enough to find used hoophouses for a fraction of their original price. Also, we didn't buy our rotary harrow and flame weeder until several years into our operation. When we began, we committed to producing 30 CSA shares in our first season and 50 in our second. At that time, we did all our harvesting on the morning of delivery day, which saved us the problem of having to refrigerate our vegetables. Later, when we upped our production to 100 shares and had to take a whole day to harvest them, a cold room became necessary.

Having said that, even though certain tools on this list may not be absolutely necessary in your first season, they do make the work much more efficient and pay for themselves quickly. This is why we have never shied away from trying out new equipment. When we first began, we seeded by hand all the crops that do not transplant well (e.g., carrots, radish, mesclun mix). Those were

Start-up Costs for Market Gardening

1 greenhouse (25' × 100')	$11,000
Two-wheel tractor and accessories	$8,500
2 hoophouses (15' × 100')	$7,000
Cold room	$4,000
Irrigation system	$3,000
Furnace	$1,150
Flame weeder	$600
Indoor seeding equipment	$600
Hoes and wheel hoe	$600
Broadfork	$200
Seeders	$300
Rakes, shovels, spades, wheelbarrow, etc.	$200
Harvest cart	$350
Floating row cover, anti-insect netting, and hoops	$600
Sprayer	$100
Harvest baskets, scales, other equipment	$300
Electric fencing	$500
Total	**$39,000**

long jobs. But when we started using the seeders described later in this book, we were able to seed beds two or three times longer in one fifth of the time. When you stop to consider the extra workload required the first few seasons, it makes sense to make the optimization of operations a priority. In my opinion, it is best not to hold off too long on getting the right equipment for the job.

In most countries, there are different kinds of government assistance programs in the form of loans and grants for new farmers that can help out with the financing for agricultural equipment. We were lucky to have received financial aid when we started Les Jardins de la Grelinette. With this kind of added support, the chances of succeeding at market gardening will greatly improve. But grants or no grants, one fact remains: keeping costs low when starting a business reduces financial risk and ensures profitability over the short term. This is a winning business model in and of itself.

Minimizing Production Costs

Revenue minus expenses equals profit. This simple equation must always be kept in mind. Obviously no one goes into farming to get rich, but one should always aim for profitability when starting a farm. Having a profitable operation spares you from daily financial stresses, prevents you from needing to find off-farm employment during the winter, and allows you to set aside money for retirement. (Yes, this is possible with a micro-farm.) Profit is ultimately what keeps the operation sustainable. Many people get into organic farming for philosophical reasons or as part of a search for meaning, but at the end of the day, market gardening is a business, and it is important to treat it like one.

Most vegetable growers today increase business revenue by upping production and sales in order to see a return on the cost of their equipment. *Scaling-up your operation* has become a popular topic at conferences and in magazines on organic market farming. But when operating a market garden, one needs to look at economics from a different point of view. While there are many kinds of ways to maximize the amount of land under cultivation when mechanized, this is not the case when using the tools and techniques described in this handbook. The production model itself is the limiting factor. So returning to the equation above, if the revenue is finite and you still want profit to be high, this means expenses must be low. This is the logic that market gardeners should follow: keep operating at low cost.

Reducing start-up costs is a good first step. Avoiding mechanization and machinery-related costs (purchase, fuel, maintenance, etc.) is another one. But the most important step of all is limiting dependence on outside labor, which generally accounts for 50% of the production costs of a diversified market farm.* In a market garden such as ours, the bulk of the work is usually done by the owner-operators with the help of one or two seasonal workers, depending on the area under cultivation and the number of greenhouses. The major operating costs are thus reduced to inputs (amendments, seeds, plant protection products), which are generally quite minimal.

In the last 15 years, Lynn Byczynski, the editor of the American magazine *Growing for Market*, has had the chance to meet with many small-scale vegetable growers. In her book *Market Farming Success*, she discussed the potential revenues of market gardening and found that the net profit margin of most of these farmers is about 50%. This means that if the total sales revenue is $80,000, about one half goes to operating costs, including

* In 2005, Équiterre released a study of the production costs on various farms that used the CSA approach. The study report is very helpful when it comes to writing a business plan and can be found in the Bibliography section of this book.

external labor and fixed costs. She points out that while the 50% margin depends on many factors, it is still relatively consistent regardless of farm gross sales. This percentage is in line with the figures on our farm and is very telling of how profitable market gardening can be. It goes to show that it is possible to maintain high productivity with little in the way of costs.

Direct Selling

Direct selling of local products is at the heart of today's renaissance of non-industrial-scale farming. Essentially, it allows producers to recover part of the profit commonly scooped up by distributors and wholesalers. Most grocery stores or food markets take a cut of between 35% and 50% of the selling price. The distributor, which transports and handles the product, takes another 15% to 25%. So, for a salad that sells for $2 in the store, the vegetable grower selling through conventional distribution channels makes about $0.65. This effectively means that if this grower doesn't participate in selling, he or she is missing out on two thirds of the value of his product—a sizable chunk. By comparison, market farmers who use direct selling channels make the full amount with every sale. We can conclude that these producers can afford to produce one third as much volume and still earn the same income.

There are several forms of direct selling (also known as short supply chains). Examples include community-supported agriculture (CSA), farmers' markets, solidarity markets, and farmgate sales. Vegetable growers who are just getting established in farming should consider these niches

Advantages of the CSA Model

GUARANTEED SALES: The main advantage of the CSA model is that production is prepaid at the start of the season, often before the first seed has been sown. This model allows the farmer to budget with greater precision. There is nothing better for a solid business plan than guaranteed sales.

SIMPLER PRODUCTION PLANS: Since members have already purchased the produce, the farmer can plan production based on the sales. Once the number of customers has been determined, the contents of each share can be planned out beforehand. This is all the more important for growers who do not yet have much farming experience to go on.

RISK SHARING: The idea behind CSA is that the risks inherent to agriculture are shared between the family farmer and the members. When members sign up, they sign a contract inviting them to be tolerant in case of hail, drought, or any other natural catastrophe. If the season is good, the members will receive more than planned, but if the season is bad, they will receive less. It is like taking out an insurance plan on the harvest.

CUSTOMER LOYALTY: CSA allows farmers to build not just customer loyalty but tangible relationships between consumers and the farm. On our farm, many members have been receiving vegetables from us for many years now. These people know us, have come out to visit the gardens, and greatly appreciate the work we do. As its name suggests, CSA really does have the power to build community.

NETWORKING: CSA is even more advantageous when a third organization can play a coordinating role. This is the case in Quebec, where Équiterre promotes CSA through publicity campaigns and finds members for the farms through its network. In addition, Équiterre provides training on production planning for new farmers, links them with more experienced growers through mentorships, and organizes visits to other operations. These are very helpful and useful services for any beginning vegetable farmer.

For more information visit equiterre.org

The development of farmers' markets and CSA are a sign that citizens are taking back the agricultural economy. Once people get a taste for real food, most don't want to rely on supermarkets anymore. This creates a lot of opportunity for new farmers.

if they hope to prosper over the long term. Moreover, the work we do as farmers addresses a need felt by a growing number of people who want to support and get to know local producers. One of the benefits of direct selling is that it provides confidence to consumers by ensuring safe, nutritious, and responsibly produced food, which is not always readily available in today's globalized food system.

That being said, one could ask which mode of direct selling is better than others. This is hard to answer, since each model has its advantages and disadvantages, and each farm has its own needs. In our particular case, although we sell our produce at two farmers' markets, CSA has always been the preferred option since it guarantees sales and simplifies our production plan. In my opinion, the many advantages of CSA make it a sales outlet tailor-made for new market gardeners.

Whatever model one chooses, the point of direct selling is to build a loyal base of customers and develop an interdependent relationship with them. When it comes to customer loyalty, the quality of the products is very important. One should never neglect the importance of presentation (for instance, always washing the vegetables) and the importance of identifying your production with a distinctive logo. Another key to success with direct selling is to be welcoming and open to sharing information with people who may be—for the first time in their lives—asking questions about where their food comes from. This is why we have always felt it important to be present at market stands and drop-off points. As growers it is important that we never lose sight of the fact that small-scale production is viable today because there is a movement among consumers to support artisan producers. Putting our faces alongside the vegetables helps to make this possible.

Adding Value to the Crops

In 2012, a five-pound bag of organic carrots sold for about $6 in the grocery store ($1.20 per pound), while the same carrots in a bunch sold for $2.50 per pound. The value of the carrots more than doubled simply by leaving the leaves on to indicate freshness. This is an example of adding value to the crops. Not all vegetables grown are of equal market value, and it is wise to invest one's energies in producing the ones that command a higher price. The first step in this regard is determining which crops are the most profitable. For exploring these ideas, there are a number of resources out there for diversified vegetable growers. The book *Crop Planning for Organic Vegetable Growers*, written by Dan Brisebois and Fred Thériault, two young growers from Quebec, is one that I highly recommend.

At our farm, we went through the exercise of quantifying the value of our production by measuring not only the total sales of each crop, but also the space and time it took to grow them. We looked at space since it is a limited resource that must be used efficiently; we looked at time in order to plan the succession of crops in the same beds. The table on page 14 shows our results. Using this as a reference, we can observe, for example, that growing greenhouse cucumbers is four times as profitable as growing turnips. Or, that a bed of lettuce brings in as much as leeks, but in half the time. This practical tool makes it easy to see which crops can perform best in the market garden.

While prioritizing the most profitable crops is an important factor for deciding which ones to grow more of, there are other means of maximizing potential sales from the garden. Investigating various options and strategies is essential when competing with supermarket vegetables produced in the industrial agri-food system (where prices are sometimes very low) and with other vegetable growers selling directly (where freshness and quality are excellent).

The box on page 15 lists some of the strategies we have adopted at Les Jardins de la Grelinette. These strategies are neither original nor guaranteed to succeed on their own, but have helped our business significantly.

Since prices vary depending on quality, growing top-notch vegetables represents the greatest challenge for a beginner. But once this goal is achieved, prioritizing certain crops and finding creative ways to differentiate products will make any market garden significantly more profitable.

Learning the Craft

If you are reading this book, chances are that you are interested in market gardening as a livelihood. Whether you want live in the countryside, work

Typical Annual Sales at Les Jardins de la Grelinette

Vegetable	Total sales	Price	Number of beds per season*	Garden space	Revenue per bed	Number of days in the garden	Rank (sales)	Rank (revenue/ bed)	Profitability**
Greenhouse tomato	$35,200	$2.75/lb.	4	3%	$8,800	180	1	1	high
Mesclun mix	$15,750	$6.00/lb.	35	18%	$450	45	2	19	high
Lettuce	$9,000	$2.00/unit	18	9%	$500	50	3	15	high
Greenhouse cucumber	$8,280	$2.00/unit	6	2%	$1,380	90	4	2	high
Garlic	$6,600	$1.50/unit	8	4%	$825	90	5	5	high
Carrots (bunch)	$6,515	$2.50/unit	14	7%	$465	85	6	18	medium
Onion	$6,075	$1.50/lb.	9	4%	$675	110	7	10	medium
Pepper	$4,400	$4.00/lb.	8	4%	$550	120	8	13	medium
Broccoli	$3,900	$2.50/unit	13	7%	$300	65	9	28	low
Snow/snap peas	$3,840	$6.00/lb.	8	4%	$480	85	10	16	medium
Summer squash	$3,690	$1.50/lb.	6	3%	$615	70	11	11	medium
Green onion	$3,360	$2.00/unit	4	2%	$840	50	12	4	high
Beans	$3,280	$3.75/lb.	8	4%	$410	70	13	24	low
Spinach	$3,000	$6.00/lb.	5	3%	$600	50	14	12	medium
Beets (bunch)	$2,900	$2.50/unit	7	4%	$415	70	15	23	medium
Turnip	$2,100	$2.50/unit	4	2%	$525	50	16	14	medium
Radish	$2,000	$1.50/unit	5	3%	$450	45	17	20	medium
Cherry tomato	$1,930	$5.00/lb.	2	1%	$965	120	18	3	high
Ground cherry	$1,650	$6.00/lb.	2	1%	$825	120	19	6	medium
Swiss chard	$1,600	$2.00/unit	2	1%	$800	90	20	7	medium
Kale	$1,600	$2.00/unit	2	1%	$800	90	22	8	medium
Cauliflower	$1,600	$3.00/unit	4	2%	$400	80	21	25	low
Basil	$1,400	$20.00/lb.	2	1%	$700	120	23	9	medium
Eggplant	$1,350	$3.00/lb.	3	2%	$450	120	24	21	low
Melon	$1,225	$4.00/lb.	5	3%	$245	85	25	29	low
Leek	$1,200	$4.00/unit	3	2%	$400	150	26	26	low
Kohlrabi	$940	$1.25/unit	2	1%	$470	55	27	17	medium
Wild leek	$840	$3.00/unit	2	1%	$420	135	28	22	medium
Arugula (bunch)	$800	$2.00/unit	2	1%	$400	45	29	27	medium
Total	**$136,025**		**193**	**100%**					

* All beds are 100' long.

** Profitability is based on a coefficient that takes into account total sales, yield per bed, and the number of growing days in the garden.

Note: The figures in this table are based on data gathered over a number of seasons. They were calculated based on the allocation of our sales (65% CSA and 35% market) and the large amount of mesclun mix we sell to retailers. They give a good indication of the most—and least—profitable vegetables to grow.

with the rhythms of the seasons, or have a more down-to-earth lifestyle, the farming vocation can be an attractive one. However, as accessible as market gardening is, growing more than forty different kinds of vegetables intensively does require know-how and a work ethic that few other vocations demand. Market gardening, just like market farming on bigger farm, is tough work, and proper training is recommended.

The best advice that I can offer someone interested in learning the craft is to get first-hand experience on an established mixed vegetable

Our Strategies for Commanding Good Prices

- We focus on the quality and freshness of our vegetables.

- We favor root vegetables that can be sold with their leaves, demonstrating that the crops are fresh.

- We avoid storage vegetables (potatoes, parsnips, winter squash, rutabagas, etc.), which for the most part take up space in the garden for a long time and cannot be marketed as fresh. We have developed expertise in two crops that we deem most profitable: mesclun and greenhouse tomatoes, which we distribute through restaurants, a local grocer, as well as our direct selling channels.

- We choose the tastiest cultivars (different varieties of the same vegetable), since we want to encourage our members and customers to discover new tastes.

- We regularly try different or unusual cultivars in order to keep our members and customers interested.

- We supplement our production with vegetables purchased from producers who specialize in crops that we have chosen not to grow.

- We force our early-season crops in order to be the first to offer them at market.

- We change our prices as little as possible, and explain to our customers and members the negative effects of "dumping" that drives grocery store prices down.

- We always wash our vegetables and display them neatly.

- We guarantee satisfaction with our products at all times, no questions asked.

- We have taken the time to design an eye-catching logo that clearly identifies our products. At the local grocery store, customers swear by our products, which they recognize easily. They know that they are supporting the farm down the road.

operation. No matter what size it is, your effort will let you see for yourself the joys and pains of the trade. No school or book can replace the experience of growing food for a season and taking in—often subconsciously—all the practices another vegetable farmer does well (or less well). In this regard, it's very important to work on a farm where the farmer is seriously interested in passing his experience on to others. From what I have observed, I believe it takes commitment to at least one full season to know if you are cut out for this kind of work and lifestyle.

That being said, there is no substitute for the experience one gains by working for oneself. This is why, after spending some time on someone else's farm, I would suggest beginning your own project. Market gardening allows the chance to get going little by little. One can start without much in the way of investment, gradually expanding the size of the plot as confidence and skills increase. Starting a CSA of 30 baskets is not such a big thing, especially given that most of the families can be friends and acquaintances. Bringing what you have grown to a nearby farmers' market is also an option. This can be done on a part-time basis and involves less commitment. We should not forget that sixty years ago most people *were* growing their own food, with some selling their extra at markets. Contrary to what some people may imagine, the farming vocation is full of rich experiences and interesting people. At the risk of repeating myself, I can definitely say that anyone prepared to invest their time to learn how to grow great vegetables efficiently can succeed in market gardening.

Finding the Right Site

It is true that plots without any faults are rather rare; but it is also true that there may perhaps be no plot among those for sale that cannot, with intelligent work, provide enough for a large family to live comfortably... A piece of land is worth as much as the person farming it.

— Author unknown, *Le Livre du Colon*, 1902

FINDING THE RIGHT SITE to grow vegetables is the most important first step for establishing a successful market garden. Soil fertility, climate, orientation, potential clientele, and infrastructure are all key considerations before investing in a site. Every site has its unique characteristics, and since there is no such thing as a perfect site, it is very important to understand and prioritize these salient points. Choosing a site for the wrong reasons can make the work of a market gardener much harder—for a very long time. For example, aspiring farmers often fall under the spell of bucolic landscapes and spectacular views, not paying attention to topographic characteristics that are essential for optimal vegetable production. Some look to buy cheap land in geographically secluded areas without realizing the implications of being three hours away from a potential market. Buying farmland because it is pleasing to the eye or to the pocketbook is one of the great pitfalls for begin-

ning farmers. Of course, various non-agricultural factors will inevitably influence one's decision: access to family land, wanting to live near family or close to a lively community, and so on. But once these personal considerations have been weighed, choosing the best growing and operating conditions is of utmost importance. The best way to assess a site's potential is to work with a checklist (see the following page). I cannot recommend this enough as it forces you to critically and systematically reflect upon a highly emotional decision. It is also advisable not to jump on the first piece of land that you visit, but rather to take the time to investigate multiple sites before making a decision. These visits could start while working on another farm, or even when the reality of starting up a market garden is still a few years away. Time spent prospecting for land is never wasted as it helps develop better judgment about site evaluation.

Site Evaluation Checklist

- Determine which hardiness zone the site is located in: Consider the implications of running a market garden in a zone that is harsher than others that may be available.

- Determine the last spring and first fall frost dates.

- Calculate the number of frost-free days in the region to determine the length of the CSA season (for instance, there are currently 150 frost-free days an hour south of Montreal).

- Determine the earliest outside planting dates for early crops, as well as a feasible date for the first CSA and/or market deliveries.

- Is there a good customer base for organic and local products in the area (restaurants, potential CSA members, farmers' markets, etc.)? Do other small producers already saturate the market, or is there room for new growers?

- How far is the site from a central market? Estimate how much travel time this represents each week.

- Is the plot large enough to meet the needs of a market garden? Too small a plot will be limiting, but too large a plot might require unnecessary investment, both in capital cost and time.

- What is the soil type? Clay, sand, or silt?

- Determine the orientation and the slope of the site: Are they favorable?

- Determine whether there are any topographic depressions on the site, and, if so, can they easily be filled, leveled, or corrected with drainage tiles?

- Determine whether the height of the water table will pose a problem for soil drainage. Is underground drainage needed? If so, estimate the costs involved.

- Will the site have enough unpolluted water for proper irrigation of the crops?

- If a water reservoir needs to be dug, can the proper permits be easily obtained? Estimate the cost of such a project with a local and trusted contractor.

- Does the site have a usable building? Does it need renovations? Is it close to the future garden?

- Is purchasing an existing building a better investment than constructing a new one tailored to the needs of a market garden?

- Does the site have access to electricity and a source of potable water?

- Is the site accessible by vehicles in all seasons?

- Are conventional crops grown on the neighboring property? If so, how will the garden be protected from contamination?

- Is the soil healthy, or has it been contaminated? Do you have evidence?

One caution about renting land: always do so with a written detailed agreement in hand. Then, in the event of a dispute over interpretation, even a small one, you will always have something to refer back to.

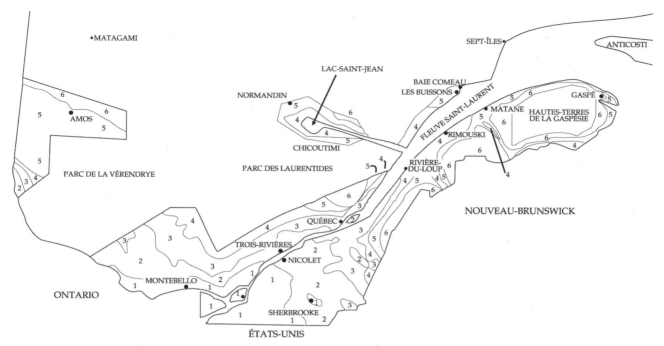

One quick look at this bioclimatic zone map of Quebec (where 1 corresponds to the warmest zone and 6 to the coldest) enables us to see that not all Zone 2 land lies in the south of the province, as one would expect. Pockets where climatic conditions are just as beneficial for growing vegetables can be found at l'Île d'Orléans, parts of Beauce, and even parts of Abitibi—way up in the north! These zones are not to be confused with plant hardiness zones.
— Roger Doucet. *La science agricole: climat, sols et production végétale du Québec.* Austin: Éditions Berger, 1994.

Climate and Microclimate

While there are many strategies that can be used to naturally extend the growing season (row covers, hoophouses, etc.), the regional climate of any site is the determining factor influencing crop growth. The number of frost-free days and the average temperature regulate both the length of the growing season and the production potential. For optimum growing potential, finding a site with the best climatic conditions is imperative.

In Quebec, there are *agro-climatic* maps that indicate the climate in different areas and its influence on the crops grown there. These maps show crop heat units (CHU), hardiness zones, and can point out the regions where one is most likely to have success growing crops commercially. Another helpful tool is the *bioclimatic* zone map, presented above, which is based on phenological data. On this map, each area is assigned a zone according to different conditions affecting plant growth, such as soil type, elevation, proximity to major bodies of water, or a specific topography—in other words, different microclimates that affect plant growth. Maps like these are a great way to investigate potential areas where local climates offer much better growing conditions.

Market Access

When searching for a potential site, it is important to keep in mind that market gardening is as much about selling as it is growing. Enthusiastic consumers ready to pay a premium for fresh, organic produce are usually found in large urban areas, but many people living in small towns also embrace the concept of locally grown, organic food. In certain rural areas, however, some folks might not be willing to pay a higher price for their vegetables (many of them grow their own), and this can become an important limiting factor for potential sales outlets. Some organic growers find that no matter how much they grow, demand always exceeds supply; others find it difficult (and frustrating) to sell their products. Location in relation to markets is critical.

It is also very important to make sure that targeted markets are not already saturated by other organic growers. Finding out how quickly demand is growing, what others are charging .for similar products, and what vegetables might be in short supply or simply unavailable are all considerations that are part of any proper market research. Taking the time to snoop around, ask questions, and identify your market niche is valuable homework.

Having the farm close to the market is also very important. Unlike growing crops, vegetable delivery requires no expertise or special attention. An hour spent on the road is an hour not spent maintaining the garden and ensuring a great harvest. This consideration is especially important when planning to sell at a farmers' market, as leaving at 4:00 AM on market day can lead to a tired and irritable farmer by the season's end. Locating the market garden as close as possible to a central market is a good investment, even if it means paying more for the property in question. As an example, our farm is located just one hour away from Montreal, but we sell 40% of our products locally at the grocery store, restaurants, and a farmers' market. This allows us to minimize our time spent off the farm, and makes us well-known and appreciated in the community.

Growing Space Needed

How big a growing area is required for the market garden is a good question. This book is geared toward growing crops on less than 2 acres of intensively cultivated land, which I believe is the optimal land base for farming without a tractor. But to answer this important question with more precision, it is important to determine how many people will be involved in the day-to-day operations and what the target revenue will be.

At Les Jardins de la Grelinette, we grow on 1½ acres (including one greenhouse and two hoophouses), and, in addition to my wife and myself, we require one full-time worker and another part-time employee to carry out the workload. I should stress, however, that we are two experienced growers working full time, year-round (in season) in the gardens. Based on our own experience and what I have seen on other intensively cultivated small acreage farms, I will advise anyone that cultivating one full acre of diverse vegetables is a lot of work for one person alone to manage. To do so successfully, outside labor will have to be hired. These people may be in the form of interns or woofers, but, if so, providing room and board will have to be dealt with beforehand.

Another helpful pointer regarding garden space required is the number of CSA shares per acre. This is often how CSA farmers relate to one another when describing the size of their farms. I've mentioned earlier that our market garden provides more than 200 families with the produce we grow onsite, and my estimates are that, in an intensive cropping system, the ratio of produce to land for a 20-week program can vary roughly between 30 to 70 shares per ½ acre of garden space. This variation depends on the farmer's experience, on how well-refined the crop planning is, and how well-designed the production systems are. These ratios are very approximate, but they do give some idea of the growing space needed to operate a market garden.

One belief I have regarding acreage is that bigger is not always better. People often romanticize about owning lots of land, but I happen to disagree with this characterization. Successful market gardening requires a lot of energy and focus. Having more than 2 vacant acres may afford extra space for raising animals, tending an orchard, and growing berries, but these additional ventures require extra planning and labor. Considering that time is often a limited resource, especially during a short growing season, one needs to be cautious about taking on additional responsibilities. The cost of acquiring extra acreage may also serve as an undesirable financial burden.

My point here is not to disenchant anyone, nor to say that owning 20 acres is a bad thing, but I find that the more land is spread out, the harder it gets to manage it appropriately. On this point, some veteran organic farmers will argue that having extra farmland allows for letting the soil lay fallow. Such agricultural practice is certainly sound, but it does require fields to be plowed and heavily tilled with a tractor. Acquiring a tractor for this purpose might lead you to start growing in a more extensive manner, thereby losing all the advantages of non-mechanized production. As tempting as it may be to branch out, and at risk of repeating myself, small is profitable.

Soil Quality

In organic growing, yields are largely dependent on the quality of the soil that nourishes the plants. The ideal soil is loose, drains well, and is high in nutritional content needed for growing healthy vegetables. You can make just about any soil fertile by building it up with proper amendments (see chapter 6), but the time and energy required to do so will be dictated by its initial quality. For this reason, it is important to know the kind of soil a potential site has to offer. This is especially important when starting a market garden on rented land, with the intention of eventually moving to another site (hopefully a permanent one). In this case, making long-term investments to improve the soil is less advantageous and aiming to find the best soil possible is essential.

10% TO 30% CLAY

30% TO 50% SILT

25% TO 50% SAND

Soil is rarely ever pure clay or pure sand but rather a mix of different-sized particles (clay, silt, sand, gravel, etc.). To determine the proportions that are present in your soil, put two inches of soil in a Mason jar and fill the rest with water. Add a teaspoon of dish detergent; the detergent acts as a surfactant that helps separate out the different soil particles. Shake the jar well and let it sit for a day. The separation of the soil into layers of different thicknesses will tell you the dominant characteristics of your soil.

Quality of soil is mostly determined by its type (clay, sand, or loam) and its percentage of organic matter. The latter can be managed and improved, but the former will greatly influence growing practices and limitations. There are a number of simple techniques for determining soil types (see below) without resorting to a laboratory soil test. But when close to choosing a site (or when not sure about different alternative choices), I recommend contacting your local agriculture extension agent for proper soil sampling and analysis. Soil tests will not only provide a better understanding of the soil type, but also of the soil organic matter content, its pH, and its chemical balance, all of which will need to be assessed at one point or the other.

The soil where we currently grow is a loamy soil, but before getting established in St-Armand, I had the opportunity to spend a couple of years working on other farms, experiencing different soil types. Having grown both on relatively poor sandy soil and in heavy clay soil has made me really appreciate what I have now. Starting with the best possible soil is a smart investment.

Topography

Contrary to what one might imagine, the perfect site for a market garden would not be on flat land, but rather on a gentle and steady south-facing slope with no depressions. Topography affects how well the soil drains, how quickly it warms up, and how fast it can be planted. And because topography is an unchangeable feature of a site, it needs to be considered carefully.

Implications of Different Soil Textures for Market Gardening

If the soil is sticky, forms a malleable ball when handled, and is hard, crusty, cracked, and generally difficult to fork, you have **clay soil.** This soil type is generally the hardest to work with, especially in the springtime, as it is slow to drain and dry out. While clay soil is rich in nutrients, it compacts easily and is susceptible to poor aeration. You may need to add perlite, coarse sand, or fine gravel to your holes when transplanting to avoid these problems. Raise your beds extra high to encourage proper drainage. Always use a broadfork before seeding or transplanting in order to improve aeration. Work the beds up in the fall to prepare for the first spring seeding, but take care not to leave the soil bare over the winter.

The structure of clay soil can be improved by repeatedly incorporating very large quantities of organic and mineral matter such as compost, peat moss, and coarse sand. This may take a number of years and require a financial investment, but when stuck with heavy soil, it might definitely be worth it.

If the soil is granular and crumbly, does not stay in a ball when wet, breaks up easily, and contains many stones or gravel, you have **sandy soil.** Difficult to compact, this soil type is generally very permeable and offers greater aeration for the crops. However, it tends to be dry, leaches water easily, and generally lacks fertility. You will need to set up irrigation for the whole site quickly, as very young plants in this soil will die after a few days without rain. To avoid leaching, make sure fertilizer is added in small doses over the course of the season. Develop a green manure program in which crop residues and compost are mixed in to increase the organic matter in your soil.

If the soil is fluffy and forms a ball that breaks apart easily, you have **loam.** Loam contains roughly equal proportions of sand, silt, and clay and brings together the ideal soil qualities: good water and nutrient retention, as well as proper drainage and aeration. Sandy loam is considered the best soil for vegetable growing. You will have to maintain the soil fertility with an appropriate fertilizing plan and conserve proper soil structure with minimal tillage.

A gentle slope (less than 5% in order to limit erosion) is a precious asset come springtime. When the snow cover starts melting, the incline in the terrain (along with the raised beds) will allow excess water to be channeled away from the growing area. The same is true during episodes of excessive rainfall, which can be quite devastating in the garden. Unfortunately, as a result of climate change, flash floods are bound to occur more frequently. The market garden needs to be well prepared to counter these inundations.

The *aspect* of the slope also affects conditions on the site, due to the amount of direct sunlight it receives. A south-facing slope means that the

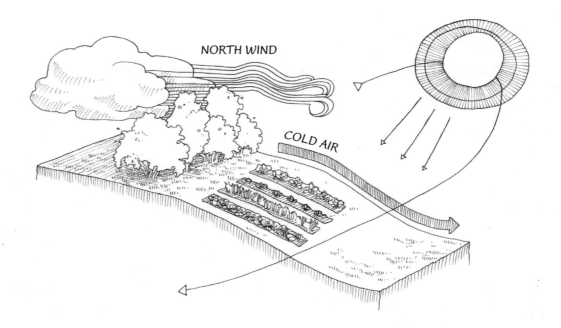

NORTH WIND

COLD AIR

A market garden located on a natural south or southeast facing slope has many advantages, one of them being that the sun's rays will hit the soil surface more directly, thereby increasing their warming effect in the spring.

maximum day temperature will be reached in the morning (in the afternoon for a west-facing slope). Gardens oriented to the south or southeast will hence warm up more quickly during each day, but also in the season. A faster-drying soil in the spring means earlier crops, which in turn might give a market gardener a considerable competitive advantage over other vegetable producers at the market. To avoid the opposite situation, it is best not to choose a site with a north-facing slope.

The steepness of the slope, along with the location of the plot, will also play a major role in how air circulates in the garden. Since cold air is heavier than hot air, it flows downward, creating natural ventilation that stirs up any stagnant air that would otherwise predispose crops to fungal diseases. This convection will also help avoid frost damage early in the season. For this reason, a market garden should never be located at the bottom of a slope, hill, or valley where it would feel the effects of frost much earlier than if it were situated on the upper half a slope.

Drainage

In Quebec's northern climate, the snowmelt and abundance of spring rainfall inevitably means excess water in the garden that must be dealt with.

Poor soil drainage can be a major problem for optimum plant growth, but even more so, it can very often prevent access to the garden when pressing work needs to be done. Growing crops on raised beds helps in this regard, but proper field drainage is still important. As discussed earlier, locating the garden at the top of a gentle slope is the best way to manage surface runoff. Digging channels that lead the water to a ditch or reservoir can then quickly divert excess water. In our garden, the orientation of the beds (east-west or north-south) was designed with this in mind. On sites where this perfect situation is not a possibility, even more attention must be given to proper field drainage.

When visiting a potential site, one of the first things to look for are spots where water pools. If such depressions are present, it must be determined whether they can be corrected, and at what cost. Certain sites will have uneven surfaces, and it is not always easy to make out slope orientation and grade by the naked eye. The way to do this is to closely watch the terrain during a heavy rain and observe the flow of the runoff. Returning to the site a few days later will allow for more and more accurate observations about wet areas.

Wherever soil tends to stay waterlogged more than elsewhere, there is cause for concern. The best solution to this is to simply avoid growing crops in those areas. Another option is to try and correct the landform by filling the depression with heavy earthmoving machinery, but this solution can be expensive and can damage the soil. Another approach is to construct surface inlets to subsurface drainage tiles. These are rather simple to construct and might just do the trick.

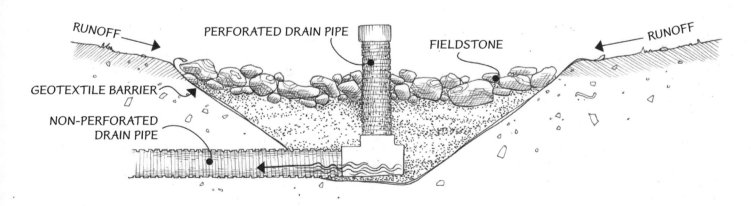

RUNOFF   PERFORATED DRAIN PIPE   FIELDSTONE   RUNOFF

GEOTEXTILE BARRIER

NON-PERFORATED
DRAIN PIPE

The most common type of "tile" is a corrugated plastic pipe with small perforations to allow water entry. Generally they are installed 2 to 4 feet below the soil surface so that water flowing into the tubing runs on a downward slope towards an outlet.

On sites where the water table sits relatively high all season, long drainage tiles will need to be installed. In order to tell whether drainage is required for your site, dig a few holes in the areas that seem the wettest and measure how high the water is in relation to the level of the soil. If the height of the water table is less than three feet from the surface of the soil, either early or late in the season, underground drainage is recommended.

Installing underground tile drainage is not a straightforward process. Special slope calculations must be made to ensure proper grade, spacing, and depth of the pipe. I strongly recommend enlisting the services of a professional for this work. Considering all that is involved in digging up a garden a second time—when the soil biology and structure have taken years to develop—mistakes should be avoided at all costs.

Access to Water

Intensive vegetable production is highly dependent on a consistent and sufficient supply of water. In our Northeastern climate, rainfall during the growing season is unpredictable, erratic, and often insufficient. A successful market garden therefore requires an irrigation system which will ensure water is always available for proper germination of direct-seeded crops, as well as providing adequate soil moisture for transplanted crops. Irrigation is also essential for providing needed water to the crops in times of drought. A potential site must therefore have a water source that meets the needs and size of the operation.

A proper well may suffice for a small home garden, but when more than one acre is under cultivation, this alone will not be enough. A water reservoir in the form of a pond, lake, or river will be needed. If a potential site already has one of these reservoirs, it is important to figure out if the water volume and recharge rate are sufficient for irrigation needs. Figuring out this information is not obvious, and the best way to go about it is to contact an irrigation equipment supplier. A reliable company should provide this service in exchange for purchasing irrigation equipment from them.

If the site has no pond or lake, digging a reservoir is inevitable. Although this task is not very complicated, it requires planning ahead. First, it is advisable to contact local authorities to see if special permits are required. In Quebec, permission is usually granted, provided that the work does not affect an existing source of water (e.g., a creek). When renting land, a letter of permission, including a detailed work plan, should be obtained from the owner. Digging a pond involves reworking a massive amount of dirt, and since some owners might not fully grasp the scope of such a project, the details of the project should be clear.

Of course, careful budgeting is important as it will determine the size and scope of the water reservoir. Surprisingly, the excavation cost of such a big project can be quite reasonable. Knowledgeable contractors can be of great service in this regard, providing they are experienced with earth ponds. I recommend visiting some of their former worksites and hiring someone who has one of the largest available backhoes. With their help, it should be easier to assess the water-retaining ability of the hole, and steps needed if the subsoil

Tim Matson's book Earth Ponds *is a valuable source of information for setting up a natural lake or pond.*

In 2012, we hired a backhoe with a large shovel for approximately $125 per hour to complete the 18-hour job of digging a pond and landscaping around it.

layers are too permeable. Finally, it's important to plan ahead for what to do with the dug up soil, as transporting the soil off-site can easily double the costs. Having a huge pile of dirt onsite might prove to be useful, as it can be used to fill in uneven ground or to raise surfaces for greenhouses and future buildings. Some of it can also be used in the garden, and for this reason, it is important to remind the backhoe operator to separate the topsoil during excavation. If the digging is done in the spring, the excavated soil might be too wet to work with, so consider carrying out this work in the summer.

Once the water reservoir has been dug, all that's left to do is to put in aquatic plants and a filtering marsh in order to transform the pond into a natural swimming hole. This oasis of biodiversity is sure to be popular not only with the birds, frogs, and salamanders, but with the kids too. Nothing beats a dip in a cool pond during the scorching heat of summer.

Infrastructure

In addition to garden space, a market garden needs a building, driveway, electricity, and water. When looking for the ideal site, there are a number of possible infrastructure scenarios, each involving a different amount of planning. These features are where big investments might have to be made. But even more important, the way the infrastructure is designed will have long-term implications for day-to-day operations, so it is important to consider them thoroughly.

Buildings: Different Scenarios to Consider

If the site already has a farm building with running water and electricity, you may well be able to get established quickly and affordably. Old barns are often ideal. But be warned that even an ideal building can become a serious impediment if it is located too far away from the gardens. In a day, month, year (or a lifetime!), trips between the washing station and the gardens are so frequent that an improperly designed site could end up making the overall operation highly inefficient. If this were the case, you might want to reconsider choosing such a site, especially if the building itself is costly.

No building, but you own the site. If the site does not have a building but you are establishing a permanent market garden, consider building a temporary structure immediately and then plan to construct a new building suited to your needs in the future. This situation is ideal because you have the luxury of time to truly figure out your needs and design a space that is multifunctional. For instance, a single building may have several different uses including temporary housing for future employees, a sales stand, a germination room, a washing station, a tool shop, and more. Before going forward, take time to visit other growers' facilities to benefit from their ideas and experiences. I also suggest prioritizing the practical concerns of the building over any other aesthetic considerations. Building a simple structure with proven techniques and locally available materials is the best way to avoid unfortunate surprises with regard to costs and timelines.

If you are renting a site without a building, there are a few different movable shelters you can consider: a prospector's tent, a yurt, or a carport. These low-tech solutions are affordable and convenient, but if there is no electricity or potable water for drinking and washing vegetables, you may want to reconsider the viability of such sites. Portable water tanks or pond water sterilized with a filtration system may be used, but this is not practical when the temperature drops below freezing. While temporary shelters can create precarious living circumstances, it is nonetheless possible to make use of them. This is how we began market gardening many years ago.

Putting asides considerations of sheltering the farmer and the workers, one or more buildings are necessary for handling and storing vegetables and for protecting tools and equipment. A driveway or dirt road needs to connect the main buildings to the public road, and should be serviceable at all times, even during heavy rain and snow. Getting stuck in the mud (or a freak snowstorm) just before a delivery is not an option. If the site has no driveway or road and one must be built, find out about municipal standards that apply; constructing a simple access way can involve enormous

We had the fortune (and the audacity) to start our market garden on the site of an old rabbit farm. We transformed one section of the building into a multifunctional storehouse, and after two winters of renovations, the other part became our home.

costs. Check this out carefully before considering a site with no existing vehicle access, or you might be in for quite a surprise.

Assessing Possible Pollution Problems

We all imagine ourselves gardening in a pristine environment. Unfortunately, external pollutants can be a reality in almost any area. The tragic story of the Montreal community gardens being closed as a result of heavy-metal contamination should be a lesson to us all. While farmland in the country may not have been used for industry in the past, other sources of contamination are possible. The intensive use of lead hydrogen arsenate as an insecticide in orchards was recommended until the 1970s, and old orchards may still be contaminated with this non-biodegradable and carcinogenic chemical. It's not always easy to know a site's history, so when in doubt, a chemical soil analysis should be done to check for the presence of inorganic pollutants. Better safe than sorry.

Abandoned fields that had been under conventional farming for a long time are usually sapped of their vitality due to the compacted soil from tractor tires, absence of crop rotation, and excessive use of synthetic fertilizers, pesticides,

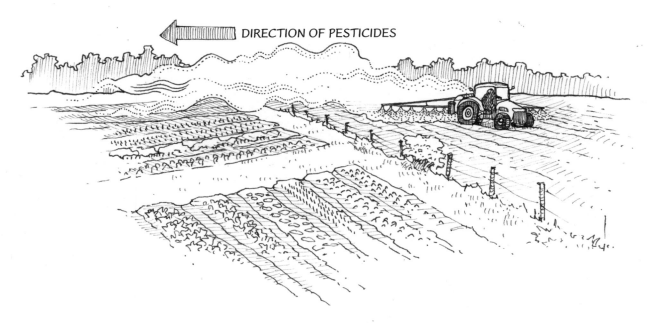

DIRECTION OF PESTICIDES

Glyphosate (Roundup) is particularly lethal to most vegetables. Unfortunately, it is the most widely used herbicide by conventional growers in Quebec.

and herbicides. Not only is this soil unhealthy, it will not be eligible for organic certification until after a three-year transition period.

Another situation for concern is when a site borders conventional farming. In southern Quebec, organic and conventional agriculture exist side by side. From the seed (coated or genetically modified with fungicides) to the growing practices (chemical fertilization and weeding with glyphosate), the results of such farming practices may well prove to be rendering toxic the water and air of nearby homesteads. But even worse, there is nothing that can be legally done to prevent these synthetic chemicals from being sprayed near the edges of an organic farm or garden. The misfortune of having crops destroyed by herbicide and pesticide drift from a neighboring farm can be catastrophic for the reputation of any organic grower.

If a desired site borders a conventional field, it is wise to take precautions. Organic certification requires that you have a 25-foot buffer zone or a windbreak between your garden and the neighboring field. It does not hurt to inform the neighbor of your intent to grow organically. The best way to persuade a conventional farmer to consider your needs while spraying is through courteous and respectful dialogue. Consider negotiating a contract stating that the other farmer will also maintain a buffer zone in exchange for some form of payment. This scenario is especially desirable if your chosen site is small.

Designing the Market Garden

There are rules to follow when designing an area. These rules are concerned with orientation, zoning, and interactions. There are whole sets of principles which govern why we put things together and why things work.

— Bill Mollison, *Introduction to Permaculture*, 1991

TAKING THE TIME to make a clear design when setting up a market garden will go a long way toward determining how efficiently many day-to-day chores will be carried out. The aim is to organize the different working spaces—inside and outside—so that the work flow will be as efficient, practical, and ergonomic as possible. The first step toward achieving this is to get a sense of all the fixed elements needed in a market garden (storage facilities, water reservoirs, greenhouses, windbreaks, etc.). Then it becomes a matter of drawing a map of the overall garden, laying out all the elements in the optimal location with regard to one another.

Buildings and Foot Traffic

Tending crops full time involves a lot of ferrying between different buildings and gardens. Without proper planning of the space, the walking distances between the tool shed, garden, washing station, and storage area can translate into serious

downtime. Moreover, visits to the bathroom, forgotten tools, or missing harvest bins are events that happen almost every day. If every trip means halting work for ten to fifteen minutes, imagine how much working time is lost over the course of a day, a week, or a season (or thirty years…). To avoid these common losses of time make sure that the immovable elements—tool shed, washing station, cold room, and bathroom—are located as close to the gardens as possible. In an ideal setting, these would all be together under one roof in a building located in the middle of all the garden plots.

The next step is to plan the working area for the washing, handling, and storing of the vegetables. This space should be as pleasant and comfortable as possible since many hours each week are spent there. Visiting other vegetable farms, regardless of the type or size, and studying how they organize their work spaces is a great way to get good ideas about how to design the best possible setup. More often than not, clever design

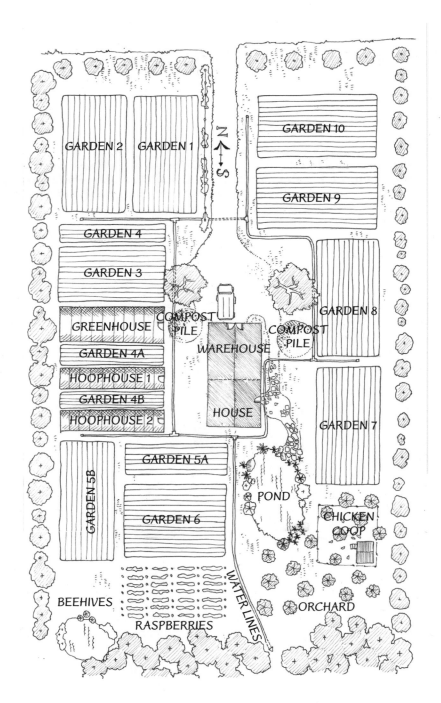

We've laid out the gardens so that all plots are equidistant from a multifunctional building that houses the washing, processing, and storage areas. This setup maximizes day-to-day productivity by minimizing onsite foot traffic.

Washing Station Setup: A Few Tips

- A simple and economical washing station consists of two tubs positioned side by side, each with its own garden hose nozzle. Ensure that there is adequate water pressure so that both nozzles can work simultaneously. The drain pipes should be designed in such a way that you can drain them at the same time without creating an overflow. The standard height for work surfaces is about 36 inches, but installing the tubs at different heights to accommodate workers of different sizes can increase comfort and efficiency. Covering the walls with waterproof materials in an indoor washing station will prevent the growth of mold. The water used for washing vegetables must be potable and drained in an environmentally responsible manner.

- Design your washing station area with enough space to accommodate a large table for weighing and bagging the vegetables. A separate area for handling deliveries and assembling CSA boxes also makes the job easier. Good shelves are needed to store bags, elastic bands, and other market equipment. Movable tables come in handy when you want to move the job outdoors in nice weather.

- Be sure to include a sink with a soap dispenser for hand washing. Sanitary norms generally require that hot water be available for hand washing.

- Lighting and windows are also very important: you don't want the station to feel dark and dreary. The floor should be smooth and easy to clean. The ideal floor is a level cement slab with one or more drains.

- Harvest containers should be easy to stack for compact storage.

- Make sure you can load the delivery truck without having to lift the harvest containers one by one. Either build the loading dock to the height of the truck bed or build a moveable loading ramp. In any case, design your space to allow for work with a dolly and containers that stack easily.

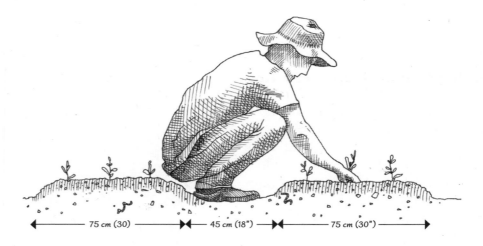

 reference placed.

The pathways in our gardens are wide enough to allow the passage of a wheelbarrow or to work in a crouching position without damaging the adjacent bed. Our beds are oriented according to the natural slope of our site to encourage surface drainage.

ideas are scale neutral, which is to say that they will work well at any scale.

Standardizing the Garden Layout

Most mixed vegetable farmers I know subdivide their fields into several smaller sized plots called "field blocks." Once the field is laid out in a compartmentalized manner, managing the complexity of growing many different crops at once becomes easier. Instead of growing ten acres, you're growing ten different one-acre plots that you can then handle individually. Standardizing the field blocks to be of equal size, shape, and length is a *very* effective way to manage different aspects of production, namely crop rotation, calculating soil amendments, and production planning.

We were fortunate to have learned the importance of standardizing growing spaces before settling permanently in Saint-Armand. In our garden, all our beds are 48 inches wide from center to center: this allows for a 30-inch band for the raised beds and an 18-inch band for pathways. The raised beds are narrow enough to step over without trampling them, and the pathways are wide enough to accommodate wheelbarrow traffic. Because more and more market gardeners are now growing on 30-inch beds, most of the new tools and equipment tailored for market gardening are developed to this standard width. When considering growing without a tractor, I highly suggest adopting a 30-inch bed system.

All our beds are 100 feet long. This length is adapted to our particular production scale and should not be considered as a standard. Other gardeners may opt for beds that are 30 feet long, 45 feet long, or any other length. The thing to keep in mind is that all beds should be of a uniform length as this renders all tarps, irrigation lines, floating row covers, and other equipment to be uniform as well. Materials cut to standardized lengths are versatile and interchangeable—and ultimately you end up needing less of them overall. Anyone who has spent time rummaging around for a piece of row cover cut to the right size will easily understand how useful this arrangement can be. Having standard lengths also allows us to treat the bed as its own unit of measure in our annual planning process, replacing the traditional per-acre calculations of yield. As a side note, we no longer make our fertilizer amendment calculations in tons per acre, but in wheelbarrows per bed.

Furthermore, our beds are grouped into equal-sized areas that we refer to interchangeably as "gardens" or "plots." The end result is that our growing space is divided into ten plots, each measuring 65 feet by 112 feet, each of them having sixteen beds that contain vegetables belonging to either the same botanical family or that have similar fertilization requirements (see chapter on fertility). The garden length of 112 feet is equal to one bed length (100 feet) plus 6 feet on either end to allow for passage of a harvest cart. Again these plot sizes and their numbers best accommodate our site and growing needs and could be different for another market garden. It's the uniformity between them that is such a great design practice.

Locating the Greenhouse and Hoophouses

Hoophouses and a greenhouse are indispensable in any market gardening operation, as they extend the growing season. Greenhouses are distinguished from hoophouses (also called tunnels) in that the latter are usually unheated (or minimally heated), have only one layer of plastic, and do not require electricity. The greenhouse serves as a plant nursery in the spring while hoophouses are used for season extension. In summer, both are used to grow lucrative heat-loving crops such as tomatoes, peppers, and cucumbers. The placement of greenhouses and hoophouses should be based on the following considerations.

Since the hoophouses and greenhouses are visited several times a day in the spring and fall to control ventilation, it is best to install them near other frequently visited facilities. In the case of a heated greenhouse, it also needs to be accessible by vehicle for fuel supply.

A north-south orientation is better for light distribution inside the buildings when the growing season is in full swing. However, an east-west orientation will capture maximum sun from September to March, when the sun is lower on the horizon. When using hoophouses for season extension, east-west is the ideal orientation.

If you plan to build structures parallel to each other, be aware that buildings may cast shadows on neighboring structures when days are short. To eliminate this problem, it is important to space the buildings one building-width apart, which incidentally is also necessary for snow removal between the hoophouses.

Protection against Deer

A market gardener should never underestimate the damage that deer are capable of doing. I have seen thousands of dollars worth of crops devoured in a single night, so do not take this lightly. If deer and other four-legged nuisances are a threat in your area, the best way to protect your growing space is to surround it with metal fencing six feet high. This solution is expensive, but effective.

When starting out a market garden or when renting land, electric fencing is a more suitable option as it is portable and more affordable. Other gardeners have often talked to me about using polypropylene mesh nets as a cheaper alternative to metal fencing, and it might be worth checking this out. As far as I know, the material is light, resistant to ultraviolet rays, and easy to install and move around. I am not sure how well it stands up to accumulated snow, however—this is something to investigate.

Finally, there is one solution that has never let us down: our faithful farm dog. Because he runs free and sleeps outside, he is a proven expert at keeping deer away from our gardens—even when we can spot over a dozen of them in our neighbor's field. Of course, barking in the night can be a nuisance, but this solution, besides providing incredible companionship, only requires a modest investment: a doghouse and some dog food.

Windbreaks

One of the most debilitating climatic features for vegetable production is heavy winds that blow continuously on the crops. Strong winds cause direct stress to plants, lower the ambient temper-

A windbreak can slow the wind speed over a distance of about 10 times its height. This creates a protected zone with more favorable climatic conditions.

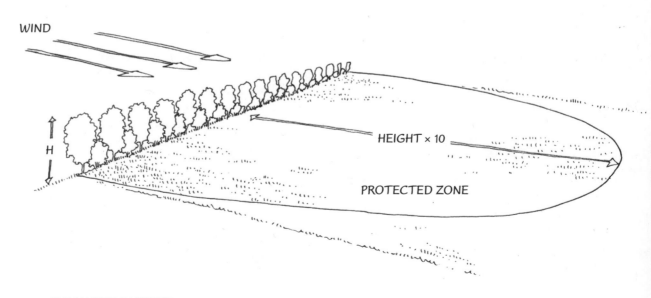

WIND

H

HEIGHT × 10

PROTECTED ZONE

ature, and dry out the soil. Given that prevailing winds almost always blow in the same direction, it is important to reduce the likelihood of damage by installing a windbreak.

A windbreak can be made of living plants (hedgerows, shrubs, or trees) or artificial materials (fencing, synthetic netting). The advantages of synthetic windbreaks are that they can be built quickly and don't take up much room. However, they are usually only six feet high and last only a few seasons. Natural windbreaks take much longer to install but are less expensive, taller, and aesthetically pleasing. Trees, shrubs, and hedges also increase biodiversity and encourage insectivorous birds and insects, which in turn reduce the number of insect pests in the garden. Specific plants around the edges of the windbreak might also attract different pollinating and predatory insects. Although these effects might be difficult to quantify, establishing natural windbreaks will add additional ecological niches to the garden habitat.

Choosing between a natural and artificial windbreak is not a mutually exclusive decision. On a newly established garden located on a windy site, it might be a good idea to implement both routes, side by side, at least temporarily, to get the best of both worlds.

Irrigation

An irrigation system is an absolute necessity for growing crops on a commercial scale. Unpredictable rainfall, rigid seeding schedules, and the need for precise crop calculations all make access to steady and abundant water essential for a successful market garden. The main purpose of

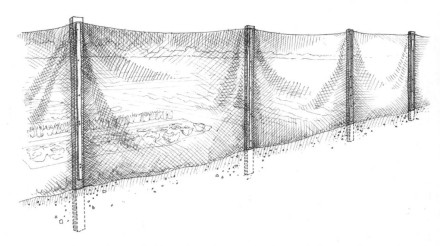

A synthetic windbreak can be an effective solution against spring gusts while your natural windbreak is still growing.

irrigation is to ensure optimal germination rates for direct-seeded crops and to provide sufficient moisture in the ground for transplanted seedlings. Irrigation can also be used for crops that require a steady supply of water and, obviously, when there is a drought. A good irrigation system should be flexible and adaptable to the specific needs of an intensive market garden.

On our farm, we favor overhead sprinklers for most of our irrigation needs. Another option, drip irrigation, uses water a lot more efficiently (since it travels slowly and directly to the targeted plants' root zone), but we find that this approach is too labour intensive. Having to remove the lines before each hoeing is a nuisance. We therefore only use drip irrigation in the greenhouse, hoophouses, and for the crops we grow under plastic mulch, where water is sure to be needed.

Since our watering needs are precise, we require sprinklers that can sprinkle in narrow bands

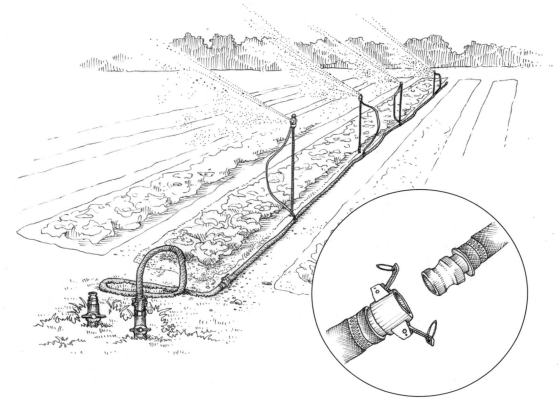

A flexible pipe with quick-connect fittings allows us to move irrigation lines to any pathway in our gardens. With two people working together, moving and installing one of them is a 10-minute job.

and opted for a low-flow sprinkler that uses a low-pressure pump (about 35 psi). These microsprinklers are lightweight, and their plastic nozzles sit on a 4-foot stainless steel rod attached to a 1-inch polyethylene (Carlon) pipe via a quick coupling. Setting them up and taking them down is quick and uncomplicated, and thus moving them from one plot to the next is quite simple. With the help of an irrigation distributor, we designed our system so that one garden (sixteen beds of one hundred feet in length) can be irrigated by two lines, each one having four sprinklers spaced twenty feet apart on the line. The nozzles of our sprinklers are set to cover a diameter of about forty feet, therefore watering eight beds at a time.

Our irrigation water comes from a pond and travels through a 2-inch main line, which surrounds our ten gardens, each of which have 2 ball valves that can supply sprinkler lines. All of our fittings are CamLock couplers (like the ones used by firefighters), which allow us to rapidly connect and disconnect the lines from the valves. Our sprinkler lines have identical couplers at each end, with connections being made by inserting a connecting plug. This enables the lines to be moved without having to point them in a specific direction.

With the help of a technician, we made sure that the pump and water lines would be the proper size to irrigate three plots at the same time (i.e., a total of six water lines running at once). Since each of the six water lines has four sprinklers, we needed to buy twenty-four of them. At maximum flow rate, all of our gardens can be watered within two days in the event of a drought. We also have two extra water lines with smaller nozzles that can spray a 16-foot diameter or, in our case, four beds at a time. One of these lines (100 feet long) has twenty-four mini-sprinklers that uniformly distribute large quantities of water in a short period of time. We use these mini-sprinklers to keep all direct-seeded beds moist. In sunny weather, we can turn them on for 10-minute stints, three or four times a day.

A word about water pumps. It's helpful to know that pumps are designed to push water over long distances but not to pull it. Therefore, it is best to install one as close to the water reservoir as possible. On our site, the water reservoir in our adjacent woodlot is located 600 feet away from our building. Although the distance is considerable, we opted for an electric pump even if it necessitated spending big money to transport electricity over such a long distance. As a result, we are able to irrigate at the flip of a switch from right inside the tool shed. That was an investment we never regretted.

I should mention the importance of installing a sediment filter on the pump. Failing to do so will result in nozzles getting clogged with small debris,

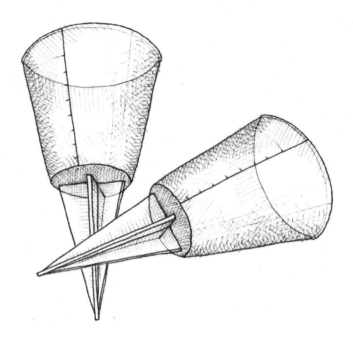

Most crops require an average of about 1¼ inches of water each week. Rain gauges are useful for measuring rainfall, ensuring uniformity, and calculating the required sprinkling time.

causing them to abruptly stop functioning. This usually happens when you're assuming that all is fine, and irrigating for long hours without paying attention.... Planning for an easily accessible discharge valve is also important. You'll need this when running just one sprinkler line, drip tapes, or when simply wanting to water with a hose connected to the main line. This relief valve will keep the main line from exploding because of excess pressure.

Minimum Tillage and Appropriate Machinery

The soil, good Quebec soil, contains within it fertile treasures. It wants nothing but to produce beautiful and plentiful harvests—but the farmer must also know how to treat it well for it to give all that it can.

— Adélard Godbout, Quebec Minister of Agriculture, *Les champs*, 1933

WHEN WE STARTED market gardening, the only tool we had for soil preparation was a rotary tiller (also referred to as a rototiller or tiller). And man did we use it! Back then it felt like the best invention ever. Just one passage with this machine would clean the soil of existing weeds and crop debris, while simultaneously preparing a smooth seedbed. The soil it worked was so loose that you could stick your whole hand into it. Eventually, however, as we became more educated about growing crops professionally and learned more about soil biology, we realized that although the tiller could rapidly prepare seedbeds, another intended result—healthy soil—was not actually happening. The tiller *seemed* to be doing all of that is desired in soil preparation, breaking up compaction and improving drainage, but its actual effect was in fact the reverse. Instead of building soil structure, we were undermining it in the long haul.

Because our first garden was on rented land, this wasn't so much of a concern; the long-term health of the soil was less a preoccupation than our short-term need for practical simplicity. However, when we moved to our permanent site, it became obvious that we needed to improve our understanding of soil structure and rethink our tillage practices. It was Eliot Coleman who first pointed us in the right direction for developing surface cultivation techniques. In *The New Organic Grower*, one of our revered books at the time, Coleman describes his different tillage practices but also suggests that reducing tillage, or even eliminating it entirely, might prove to be the ideal way of growing great crops. He asserts, however, that the main challenge to the practice of non-tillage or low-tillage is preparing soil as efficiently as conventional tilling does. We understood what he meant because growing crops intensively requires a lot of soil preparation. Organic matter must be incorporated to maintain soil fertility, seedbeds must be prepared to encourage good germination, soil must be loosened and aerated to acclimatize young transplants, and crop residues have to be disposed of in time for the next seeding. All of this involves working the soil substantially.

Currently at Les Jardins de la Grelinette, our philosophy is to try and replace mechanical tillage with biological tillage. As we see it, earthworms play such an important role in healthy soil structure—their tunnels providing aeration and drainage while their excretions bind together soil crumbs—that we want to have them thrive in our gardens. We also believe that microbes, fungi, and other soil organisms, given that we do not compromise their action with soil inversion, can perform much of the tillage needed to create and maintain loose, fertile soil. As much as this sounds good, however, we still need to intervene mechanically to prepare seedbeds for crop planting and establishment. Finding the appropriate equipment and techniques to do so efficiently, without damaging the soil structure and the living organisms within it, has been a challenge for many years.

After several seasons of experimentation, we now use a method that is biological, practical, and appropriate for commercial growing. In the middle of the season, soil preparation on our permanent beds goes as follows:

- Green manures* and crop residues are shredded with a flail mower, then covered with a black tarp for two to three weeks. This smothers the previous crop and cleans out the weeds.
- We use a broadfork to deeply aerate the soil and encourage root development in the next vegetable crop.
- We spread the amendments over the bed and mix them in with a rotary harrow to a depth of 2 inches. The rotary harrow has a roller in the back that tamps down the soil and levels the bed.
- We do a final raking to remove any remaining debris and stones. The bed is now ready to receive the crop.

Not counting the tarp treatment, it takes about 30 minutes to prepare one 100-foot-long bed. For maximum efficiency in the preparation step, all the tools we use are standardized to a width of 30 inches and can do the job completely in a single pass. The elements of our system are described below in greater detail.

Permanent Raised Beds

Raised beds form the foundation of our intensive cropping system. They provide the most space and labor-efficient layout for the market gardener and the most beneficial growing environment for the plants. The fact that they are permanent is key as it allows an optimal way of building and maintaining soil structure. After many years growing in such a system, I find it hard to even imagine growing vegetables any other way. Here is a list of the benefits of growing crops on permanent raised beds.

* If the green manure is thick, we sometimes partly turn it under with a quick pass of the power harrow or rototiller (to a depth of three to five inches) before covering it with a black tarp.

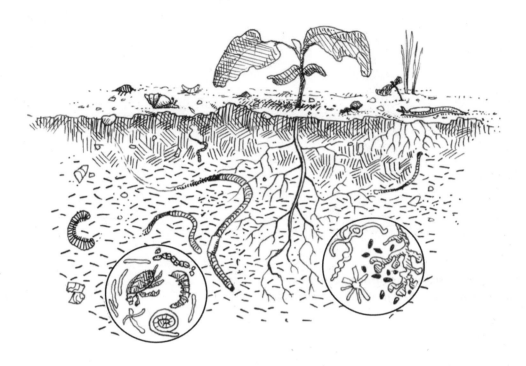

Given the right conditions, soil organisms can perform much of the tillage needed to create and maintain loose, fertile soil.

Better drainage. Raising the soil above ground allows excess rainfall to drain away from the crop zone and moisture to stay in the root zone, where it is most needed. In our northern wet climate, this practice is critical.

Soil warms faster come springtime. Since the beds are raised several inches off the ground, they intercept more of the sun's rays during early spring. Faster drying and warming soil permits earlier seeding and transplanting. Plants will also grow faster once they are established.

No soil compaction. Beds are never walked on during the growing season, let alone compacted by heavy machinery. In this system, only the pathways get compacted by the grower's footsteps. Avoiding compaction promotes loose soil structure, which in turn allows vegetable roots to extend deeply into the soil.

Higher yields. Unlike the typical single rows separated by pathways, plants in a raised bed system are uniformly spaced over the surface of a wide bed, allowing for a high plant density. In other words: increased yield per square foot of growing space.

Soil building. Using the same layout of beds and paths each year restricts organic amendments to

Depending on the size of the garden, establishing permanent beds may take a few days or even weeks.

where they are needed—the beds. Considering the large volume of amendments and compost required in an intensive system, it is the most economical approach to soil building.

Leaving out the tractor. A permanent bed system saves the work of building new beds every year and is the most efficient way of farming without a tractor. Working and shaping large quantities of soil every year would otherwise require a tractor in order to work efficiently.

For all of the reasons mentioned above, I strongly encourage beginning growers to adopt permanent beds when organizing a market garden. But this being said, note that such a set way of doing things does require initial onsite preparation.

Any major earthmoving project has to be dealt with first. Bumps and dips in the soil surface should be corrected, and if tile drainage is required, it will also need to be installed. When taking over a previously vacant site (e.g., a field or unused farmland), it might inevitably be necessary to use heavy machinery (plow, chisel, rototiller, etc.) in order to bring the land into a "workable" state. A tractor may also be needed to remove any large rocks from the site. When planning for this kind of work to be done, it might also be a good idea to establish an action plan for eliminating persistent perennial weeds such as quack grass, dandelion, and thistle. Repetitive tillage with large disks or harrows might help in that regard.

Once the groundwork is finished, then the real work begins. Depending on the size of the garden, setting up permanent beds may take a few days, or even a few weeks. Creating ours took a while since we had about 180 of them, each 100 feet long. The first thing we did was to mark off the perimeter of each plot (calculated to contain 16 beds of

48 inches center to center). We then used strings to indicate the width of each bed and dug the earth from the pathways onto the beds. It was a lot of work, but we were motivated by the fact that we would only be doing this once.

While laying out the beds, we also added large amounts of organic matter to improve the soil's quality. In our case, the initial soil was already a desirable gravelly loam soil, and we incorporated about 7 wheelbarrows per 100 foot bed of a compost mix rich in peat moss. We also added lime to raise our pH, which at the time was on the acidic side. We've seen other market gardeners add sand to clay-based soils and clay to sandy soils. Along with adding compost, these amendments help improve a soil's texture.

With regard to the height of the beds, I recommend mounding the soil about 8 inches. Over time the soil will settle, and after one or two growing seasons, the beds may be only 4 to 6 inches high. Raising the beds higher than 8 inches does not produce any significant advantages and only creates more work and higher costs. Many market gardeners seed their pathways with clover, but at our farm, we don't follow this approach. We use the soil in our pathways to pile on top of beds that have already settled, and we also use the pathway soil to weigh down tarps and floating row cover.

The Two-Wheel Tractor

A two-wheel tractor, which we invariably refer to as the walking tractor, is the ideal power source for the market garden. Versatile, robust, and easy to use, they are designed to do soil work on small cultivated areas. Although they are slowly becoming

Raised beds have a tendency to settle over time. For many years we hilled them with a plough hooked up to a small rototiller. Now we use a rotary plow fixed to our walking tractor. Every spring, we make sure to build up a few beds to ensure that they are all maintained every two to three years.

About Soil Building

Since every new garden plot varies in its initial soil texture, I would not feel comfortable giving the reader particular recommendations for soil building amendments and quantities. However, I do feel like sharing this piece of advice: with regard to building soil, don't be cheap. Remember that permanent beds are *permanent*, so investing in quality organic matter and compost will help your soil produce high-yielding and high-quality vegetables—essential components for a successful market garden. If your soil needs conditioning to improve its texture, don't hesitate to do it.

The handlebars of a two-wheel tractor can be turned sideways to allow you to walk without stepping on the newly prepared soil surface. Its light weight and small size make it an appropriate choice of machinery for cultivating in a permanent bed system.

more known in North America, walking tractors are popular in Europe, especially in Italy where some of the better-quality ones are built.

We began working with such a machine after using the small garden-sized rototiller typically found at most hardware stores. While these rototillers can provide adequate tillage, there is a world of difference between them and the rugged walking tractor, whose gear shifting and locking wheel differential make it much more powerful and maneuverable. Like a 4-wheel farm tractor, they are designed to run many attachments with a single power source. Their PTO* can run all sorts

* PTO stands for Power Take Off. This is the part where the engine of a tractor powers the implement via a rotating shaft.

of implements, such as snowblowers, generators, grass mowers, and even hay balers. Walking tractors are distributed by different companies whose experts are in the best position to offer advice regarding tractor and wheel size, horsepower, and other features. Recommended trade names and company addresses are provided in the tool appendix.

For us, the choice to use a two-wheel tractor in preference to a full-sized one was obvious from the outset. First, our site's limited growing area discouraged us from wasting space for tractor turnarounds and headlands. A walking tractor, on the other hand, can pivot on the spot, thus allowing for better use of the available land space.

Second, we found that when shopping around

for tillage equipment, the ones we favored were more readily available for the two-wheel tractor. Thirty-inch-wide implements are common for two-wheel tractors, but much less so for four-wheel tractors. Third, the price of a two-wheel tractor is much cheaper. A new one, including a flail mower and a rotary power harrow, cost us a fraction of the price of the compact tractor we were also considering. In the end, it was an easy choice that we never regretted making.

A two-wheel tractor usually comes equipped with a rear-tine tiller. As I mentioned earlier, this tool isn't good for soil structure as it pulverizes and breaks up aggregates in the soil, making it more susceptible to compaction. Nevertheless it's useful for mixing amendments and green manures into the soil and for preparing seedbeds when there is not time to cover the soil with a tarp or to manually remove crop residues. It is also helpful for really early springtime seedings when the soil is still cold. A quick pass with the tiller brings oxygen into the soil, warming it up for early production. At this time, earthworms and other soil creatures (who obviously don't like to be pulverized by tilling) aren't around, so it doesn't bother us too much to invert the soil layers of a few particular beds.

In our garden, the equipment of choice for bed preparation is a rotary power harrow (often just called a power harrow). On this tool, multiple sets of tines rotate on a vertical axis, which till the soil horizontally. The result is that soil layers are not inverted and no vertical compression of the soil can lead to hardpan formation. It works the soil by *stirring* it instead of *mixing* it like a tiller does. Our power harrow is equipped with a steel mesh

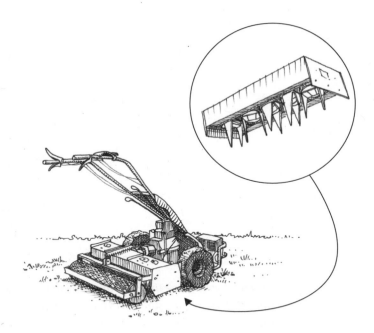

The rotary power harrow eliminates inversion of the different soil layers, which is useful in preventing dormant seeds from being worked up. Overall, it is a better bed preparation tool than a rotary tiller.

roller in the rear, which allows for incremental adjustment of the working depth of the tines. The roller also levels and pre-tamps the soil for good seed-to-soil contact. One pass with our power harrow gives us a perfectly conditioned bed for transplants and direct-seeded crops. Overall, it's an amazing tool, and I would recommend trying it first before adopting any other tilling or spading device. The only downside is its heavy weight, making it more challenging to maneuver than a rear-tine tiller.

We also have a heavy-duty flail mower that shreds green manures and crop residues with astonishing ease. It was the acquisition of this equipment (also attached to our walking tractor) that

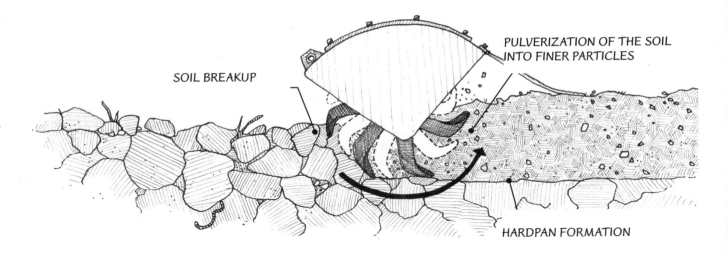

SOIL BREAKUP

PULVERIZATION OF THE SOIL
INTO FINER PARTICLES

HARDPAN FORMATION

Although a rear-tine tiller is a practical device, its use ruins the soil structure every time. Pulverising the soil leaves it nice and fluffy for a short while, but by breaking up soil aggregates, the soil will eventually settle into a more compacted state. The weight and action of the tiller also causes a hardpan below where the tines reach, further decreasing soil drainage and the ability of roots to penetrate the soil.

allowed us to pursue working with green manures in the garden. Before that we only had a regular mower, which cut the stems at the bottom of the plant, leaving us with lengthy material to work with. When we then tried to incorporate that material in the ground, the green manure would get tangled around the teeth of our tiller, all too often jamming the machine. The flail mower, on the other hand, reduces mowed material to a very small size that is easy to incorporate—the mower's many tines acting like knives to chop everything up. This was such a breakthrough in our market garden that I feel it is an absolutely necessary tool when thinking about growing green manures. The flail mower also makes it very easy to clean away old crops, a job we used to do by hand.

Choose your implements wisely as this will directly influence your ability to work the soil properly. Before making any final decision about purchasing a two-wheel tractor, consider the implements you'll be working with, as this might determine the brand, size, and engine horsepower. Not all walking tractors are equal. For example, some of them don't have the ability to reverse the handlebars, which allows the tractor front or rear PTO capability. This feature is important because implements like the power harrow perform best in the rear (so that tire tracks are eliminated) while others like the flail mowers work best in the front. It might also be interesting to consider some of the smaller tractors available from Europe and Asia, as well as those being currently developed in the

earthtoolsbcs.com is one of the best resources available about two-wheel tractors and implements.

United States. Small electric cultivating tractors are not available yet, but they should be on the market within a matter of years. No matter which machine you choose, make sure that the implements you want are available and compatible.

The Broadfork (Grelinette)

The broadfork is a long U-shaped fork that loosens the soil about one foot deep without inverting or mixing it. Because of its ability to work the subsurface of the soil, this tool is an essential part of our growing system. It complements our other tools that are all intended for surface cultivation. The way the broadfork works is quite simple. The operator steps up on the crossbar, using full bodyweight to drive the tines into the ground, then steps backward while pulling backwards on the handles, causing the tines to lever upwards through the soil. A steady worker can broadfork a whole plot (16 beds, each 100 feet long) in about two hours' time. Broadforks are also ergonomically designed to keep the user's back straight, thus preventing the common aches and pains felt after a long day of manual work.

Other growers often argue with me that using a broadfork on a commercial scale isn't efficient, but for the time being, I don't see many realistic alternatives. The broadfork is a simple and economical solution that ensures well-aerated soil. As much as I think surface tillage is one of the answers in moving towards creating biologically intelligent growing systems, for many years I have observed the beneficial effects of deep tillage with a broadfork, and I strongly feel that the advantages of this tool are too important to let laziness dictate

The broadfork is an essential and appropriate tool for the market garden given that it allows deep tillage of the soil while preserving the topsoil structure.

our practices. It is important to note that we use it only for crops whose roots benefit from deep tillage, and do not use it routinely before every single crop.

The broadfork traces its origins back to the *grelinette*, a tool invented in France by André Grelin in the 1960s. The tool we use on our farm is not the authentic French *grelinette* but rather a Canadian-made one modeled after it. We named our business, Les Jardins de la Grelinette, after the tool because it is so emblematic of our philosophy of efficient, environmentally sound, manual organic gardening.

Why worry about soil inversion? The delicate ecology of soil develops as it does for a reason. Bacteria, fungi, and earthworms, which all work to create soil structure, are found at a certain depth in the soil because it has the right moisture and aeration conditions. Turning your soil right upside down disrupts this ecology for at least a while, so that you cannot rely on natural forces to help do the job for you. Inverting the soil also brings up into the topsoil weed seeds that were dormant in the subsurface.

Tarps and Pre-Crop Ground Cover

One of the most important discoveries we made throughout the years has been that of relying on soil-covering tarps to smother crop debris when preparing new ground.

Until then, our only way of clearing the remains of finished crops and established weeds was to either till them into the ground with multiple passes of the rototiller or manually remove them. As we started to move away from relying on the tiller, we often favored handpicking out weeds and residues, rationalizing this time-consuming activity by telling ourselves that all this organic material we were bringing to the compost pile would eventually become great soil-building material. This way of working was labor-intensive and time-consuming, and our beds were never really cleaned of the smaller weeds.

Then one midsummer's day, I bought a big black UV-treated polyethylene tarp to break open new ground where we planned to plant berries. My idea was to dry it out before putting it away in the shed. As it happened, it stayed there for 3 weeks, and when we finally moved it away, it struck us—the tarps had killed all crop residues

I believe plastic tarps are just as beneficial to soil as other forms of mulching. We are often reminded of this when we pull back a tarp and are greeted by an abundance of earthworms.

and weeds, leaving us with a very clean bed surface to work on. We had stumbled onto a technique that was highly effective. Ever since, we have used tarps to cover the ground as a complement to our minimal tillage system.

Every time a harvest is done with, we immediately cover it. Depending on what crop is next in line on the schedule, the tarp will remain on the bed for anywhere between two to four weeks, leaving our minds worry-free. Passively, we are preparing the soil for the next seeding while also weeding it just like false seedbed would.

Over time we have consistently noticed a difference in weed pressure on beds that have been covered compared with others that have not. The explanation is simple: the tarp creates warm, moist conditions in which weed seeds germinate, but the young weeds are then killed by the absence of light. Looking into this, we found out that French growers were widely using this technique (called occultation) to diminish, or even eliminate, weed infestations in their fields.

To Till or Not to Till

The relationship between soil health and tillage is a hot topic for debate among practitioners and researchers in the organic farming world. It's generally understood that plowing, disking, and harrowing, although so practical for soil preparation, also have their downside. Working the soil makes it more prone to erosion, disintegrates its structure, and damages living organisms. Tillage also takes time, and remains the major use of fossil fuels on organic farms. Even some conventional agricultural scientists now claim that soil fauna

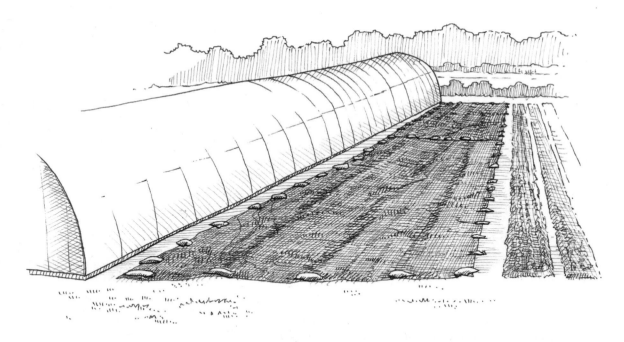

Covering the soil with an opaque tarp for a few weeks is how we manage to clean the surface of our beds without having to work the soil. This technique is also quite beneficial in reducing weed pressure for the following crop.

and flora are too easily affected by soil disturbance, concluding that tillage should be avoided. For them, however, no-till is associated with increased chemical herbicide use, such as glyphosate in seedbed preparation. For me, replacing tillage by the use of Roundup is not something I can vouch for.

The concept of no-till has been around for a long time, and many books, articles, and websites assert the benefits of not disturbing soil integrity when growing crops. The general thrust behind the concept of no-till is that amendments and crop residues are better-off left on the soil's surface than incorporated by tillage. Mulched or-

ganic matter is eventually incorporated by the action of worms, fungi, and microbes that thrive in undisturbed soil. This practice mimics the actions of forest floors, which do not require tillage and can sustain themselves for centuries, if not longer.

While in general this is a great concept, in the market garden, no-till is restrictive and somewhat impractical. Based on my experience, direct seeding into crop residues, mulch, or crimpled down cover crops is not straightforward, causing unpredictable germination rates—a nightmare for any commercial grower. Also, shallow cultivation of the soil is the best way I know to keep weeds in check and prepare seedbeds. Moreover, I have

In 2007, Eliot Coleman and his team developed a small cultivator that is operated using a battery-powered drill. The cultivator makes it possible to mix in compost and refine the soil in the top few inches. Although it is often not powerful enough to be used on the whole garden, it can be a useful tool in greenhouses, where a walking tractor doesn't maneuver easily.

never come across direct evidence that tilling the soil results in lower yields. All the best commercial growers I know heavily till the land and still obtain great results in terms of their quality of production and soil sustainability.

For us, at this point in time, we find that a middle ground is the best approach. Our current minimum tillage methods give us satisfying results when it comes to increasing yield and saving time and effort. I feel we have struck a balance between theory and practice, yet I am convinced that this is only the beginning: by keeping up to date with new ideas and strategies, I am confident that we will develop better ways to further harness the biological power of the soil, instead of relying on mechanical solutions. Relying on and working in harmony with ecological systems is the wave of the future.

In conclusion, I feel it is important to stress this final point: while many aspiring organic growers gravitate towards ideas associated with permaculture, sustainability, and alternative energy, it's also important to understand that market gardening is a livelihood before anything else. Although experimenting with novel ways of doing things is exciting, I would not be too hasty in brushing aside proven solutions from experienced growers, even if they do not seem "ideal." When first starting out, the important thing is to be successful at growing vegetables. If the idea of minimal tillage or no-till inspires you, keep in mind that it is an approach, not a doctrine.

Fertilizing Organically

IN THE PREVIOUS CHAPTER, I talked briefly about the biology of the soil and mentioned its importance in building soil health and fertility. Indeed, the fundamental principle of organic agriculture is that fertility depends upon the status of the soil (physical, biological, and chemical) and on the living organisms that make it up. Unlike the conventional approach to agriculture in which crop needs are met by fertilizing plants directly with soluble synthetic fertilizers, the organic approach recognizes that supplying plants with the nutrients needed for optimal growth is a task best left to soil micro-organisms. The best way to understand this is to imagine that life forms, such as fungi, earthworms, etc., are the "engine of the garden." The job of the organic grower is therefore to create natural fertility by adding compost, animal manure, and green manure, known collectively as organic amendments. Unlike synthetic fertilizers, these amendments must be "digested" by soil micro-organisms in order to make their nutrient reserves available. This relationship between the biological activity of the soil and organic matter management by the grower is the basis of the organic approach to fertilization.

When we started market gardening, the notion of feeding the soil and letting the soil feed the crop was comforting in its simplicity. At first, our strategy was to give the soil as much compost as we had and hope that our crops would turn out for the better. But as we learned more about the different strategies organic farmers use to fertilize their soil, we began to realize that simply piling on more compost was too simplistic. Moreover, our problem was that the amount of compost on hand was always limited. We had to find some way to better rationalize our system.

Over time, we acknowledged the importance of different tools, such as soil tests and fertilizer calculations, in order to better match the specifics of our soil to the needs of the crops we wanted to grow. In other words, we stopped fertilizing blindly. And although we already understood the importance of proper crop rotation, it took us a

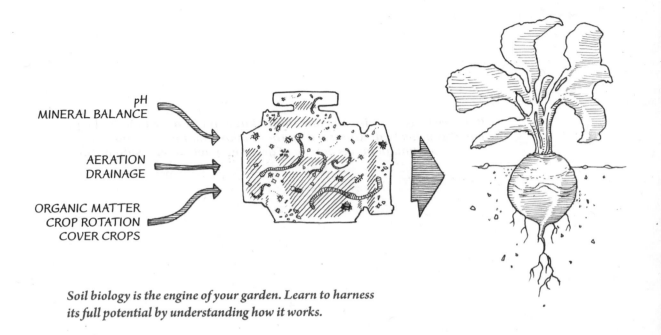

Soil biology is the engine of your garden. Learn to harness its full potential by understanding how it works.

while before we were able to use it to our advantage without it becoming overly complicated. Our many experiments with cover crops also eventually taught us how to make the most out of them in our intensive production system. All these lessons eventually led us to develop a systematic approach to fertilizing our soil and the crops we grow. For almost a decade now, we have been following a fertilization plan which has supported the growth of our intensive methods of production. The result of this activity has been more than satisfactory, especially given that our soil seems to be better now than when we started.

Looking back, learning how to fertilize organically was not a straightforward endeavor, and the assistance of agronomic expertise, especially in the first years of establishing our market garden, was of great help. Money and time spent for such expertise was an investment. I also believe that it was our interest in the biological world that allowed us to make sense of our observations in the gardens, which in turn helped us to evolve our practice into something more than simply following recommendations. All things considered, if there is one aspect of organic market gardening where lectures and studies might prove useful, I think this is it. A better understanding of what enhances soil fertility can translate into better growing practices. We see the proof of this when we harvest high yields of top-quality vegetables, at the same time as we see our gardens getting ever more productive, and with increasingly less effort.

Elements of a Proper Fertilization Strategy

- Laboratory soil testing and the correction of nutrient deficiencies or imbalances in the soil, with the help of recommendations from an agronomist.

- Proper liming to ensure the correct soil pH on all plots.

- Crop rotation plan that includes cover crops.

- Fertilizer amendments calculated on the basis of existing soil fertility and the nutrient requirements of each vegetable crop.

- Observation procedures to measure how fertilizer applications influence your crops and soil over time.

Soil Tests

As the drawing on the opposite page illustrates, the main idea behind fertilizing crops organically is to encourage biological activity in the soil. It is the micro-organisms that convert organic amendments fed to the soil into nutrients that can be taken up by plants. The agronomic term for this conversion is "mineralization." And mineralization is most effective in specific soil conditions: a balanced pH, a supply of organic matter, good mineral balance,* adequate moisture content, and enough heat to power everything. Since evaluating these conditions is impossible with the naked eye, laboratory soil test analysis is very useful—I would even say indispensable. Granted, a well-

amended garden may not show any signs of problems and can give beautiful harvests to a grower who is working intuitively. It's entirely possible to grow high yields without doing any soil testing at all. But it is also possible to enrich the soil excessively and to not realize it, leading to waste and potential pollution. Therefore, be sure to take advantage of the picture given by soil testing.

Soil sampling is simple. Prepare samples which are representative of the whole growing area by digging out dirt with a trowel to a depth of six to eight inches. Label them well and send them to a lab. When the results come back, I recommend emailing them to an agronomist who specializes in organic agriculture. Proper interpretation is key to good recommendations, so chose this agronomist wisely. You can submit your samples to any lab you want, but for good recommendations to come out of it, a thorough soil auditing should be accompanied by at least one onsite visit of the gardens by your chosen agronomist.

---

* It is common for soil to contain too much of one element relative to the others, for instance too much magnesium with respect to calcium. This imbalance can lead to poor natural fertility in the soil. An agronomist will be able to detect imbalances and propose corrective measures.

Fertilization Program at Les Jardins de la Grelinette

Here is the program we followed for many years in our intensive market garden. These recommendations are based on the needs of the different vegetables, assuming that they are grown in soil with a balanced pH and a sufficient amount of organic matter. Since our soil has become rich over the years, we have now begun reducing our compost applications compared to the program below. It should be noted that all beds in our gardens are 30 inches (75 cm) wide and 100 feet (30 m) long. You will have to adjust these ratios if you plan to implement these recommendations on plots that differ in size from ours.

HEAVY FEEDERS (Solanaceae, Cucurbitaceae, some Brassicaceae)

Granulated poultry manure	0.7 short tons/acre or 1.6 gallons/bed
Compost	36 short tons/acre or 5 wheelbarrows/bed

ONIONS (including leeks and green onions)

Granulated poultry manure	1.1 short tons/acre or 2.6 gallons/bed

LIGHT FEEDERS (root vegetables, mesclun [salad mix], lettuce, and greens)

Granulated poultry manure	0.9 short tons/acre or 2.1 gallons/bed

PEAS AND BEANS receive no fertilizer.

GARLIC is fertilized in the fall as a heavy feeder with 36 short tons/acre (5 wheelbarrows/bed) of compost. Our fertilization plan accounts for the mandatory alternation between heavy and light feeders from year to year. In our rotation, each bed in the garden is treated with compost once every two years.

If a heavy feeder comes after a leguminous green manure in the rotation, we reduce its dose of granulated poultry manure by half.

NOTE:

🐛 We have a separate fertilization plan for greenhouse tomatoes and cucumbers.

🐛 Our wheelbarrows hold 5 cu. ft. of compost and weigh about 100 pounds when full.

Crop Requirements

One of the benefits of modern agricultural science has been to teach us about the nutrient needs of the different vegetable crops. In Quebec, this information is compiled in the *Guide de référence en fertilisation*, which most agronomists will have a copy of. In the United States, state university extension services provide similar guidelines. This information, combined with soil test results and the N-P-K (nitrogen, phosphorus, potassium) of organic amendments, makes it possible to calculate fertilizer dosages for every crop grown in a specific soil. Since these calculations can get a little complicated, this is another area where the help of an agronomist may come in handy. I have to admit I find it hard to believe that mathematical equations can convey the true complexity of the many biological interactions taking place in our soil, and I have some reservations about them. Nevertheless, this practice has its merits as a guideline. The fertilization program at Les Jardins de la Grelinette is based upon such calculations.

Even so, I would not worry too much if your fertilization program is not based on such calculations. More general fertilization guidelines can be adopted and afterwards adjusted by your own observation. In the end, it is the farmer's eye that can best dictate the proper dosages for applied soil amendments. It is always good to remember that fertilizing too much is just as bad as not fertilizing enough.

Managing Soil Fertility

Like I've mentioned earlier, I now firmly believe that in order to become a good grower one needs to understand some basics about what makes a

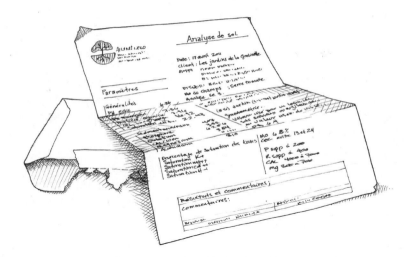

Soil testing gives you a snapshot of your current situation, and while it is not a perfect portrait, it does give you some idea of the natural fertility underfoot. It can alert you to mineral deficiencies before they appear in your crops and help you track changes in your soil over time. In my opinion, a soil test is something no farm of any size should do without.

You can think of fertilization calculations as targets. Even if they don't translate into perfect recommendations, they offer something to aim for and prevent us from fertilizing blindly.

soil fertile. In the present chapter, I'll be touching briefly on this topic, focusing on how elements affect production and what we, as growers, are able to do about it. For a more thorough elaboration on the matter, I suggest reading two instructive primers: *Building Soils for Better Crops* by Fred Magdoff and Harold Van Es, and *The Soul of Soil* by Joseph Smillie and Grace Gershuny. The references for these books can be found in the bibliography.

Organic Matter

Organic matter plays a fundamental role in soil fertility. When mineralized by organisms in the soil, it makes nitrogen, phosphorus, sulfur, and several micronutrients readily available to plants. Any organic matter not mineralized builds up in the soil and contributes to its structure. For all soil organisms, organic matter is both fuel and habitat. Biological activity and organic matter are therefore tightly linked to one another.

How much organic matter is present in the soil (in percent) is one of the main things soil testing will tell you. With this information in hand, you manage organic matter in 3 different ways:

- Build your soil by adding a large initial organic amendment to obtain a high level of organic matter in your garden. An excellent way to do this is to add peat moss.
- Maintain soil fertility by making up for lost organic matter due to mineralization, tillage, erosion, and uptake by plants. This is the main reason for using compost, green manures, and crop residues in an organic system.
- Make sure that the organic matter level in

the soil test is not high as a result of insufficient biological activity resulting from unbalanced pH or poorly drained soil—this would mean that the organic matter has just been accumulating rather than decomposing. Even if your results look good on paper at first glance, you may still have to work on the availability of organic matter in the soil by improving its physical characteristics.

pH

Most of the soils in Quebec are slightly acidic. Since a pH under 6 inhibits microbial development and diminishes soil biological activity in general, soil acidity must be corrected. This is done with liming materials such as wood ash or agricultural limestone (referred to just as limestone). The latter is more common and is what we use in our gardens. Agricultural limestone is a heavy white powder obtained from mined and crushed rock. It is a natural ingredient and acts rather slowly, which is good thing.

The ideal pH for most crops is between 6 and 7; generally speaking, you should aim for a pH of 6.5. To increase your pH using limestone, incorporate small doses gradually to avoid changing the soil chemistry too suddenly. It is good to follow agronomic recommendations on this point. It is also important to measure the pH before each application to be sure the treatment is appropriate—yet another reason why soil testing is so useful. Once you have achieved your target pH, regular additions of compost, which is often slightly basic, should be enough to keep the pH in balance all by itself. In our gardens, we broadcasted limestone on the surface and then rototilled it into the top six inches of soil.

Nitrogen (N)

The organic matter already present in the soil will release some nitrogen every year through mineralization, but not enough or not in time to meet the needs of most vegetables. Since there is a direct relationship between nitrogen supply and vegetable growth, it is up to the grower to ensure that crops get enough of this element. The added compost, manure, and other organic amendments must contain sufficient nitrogen to ensure adequate fertility. This is why all amendments are not of equal fertilizing value nor universally appropriate for each crop.

Since nitrogen encourages leaf growth in plants, it must be available immediately after crops are planted, at the time when they will be growing new leaves. When using an organic fertilization approach, we need to remember that mineralization can happen only when the soil is warm. Generally speaking, biological activity is very low or even nonexistent when the temperature is below 10°C (50°F) (in the soil, not in the air). Therefore, when the soil is still cold in the springtime, we need to compensate for the potential lack of nitrogen by adding fast-acting natural fertilizers. This ensures that the crops will get a good takeoff. Blood meal, fish meal, or pelleted chicken manure are examples of different natural fertilizers that will release nitrogen faster than compost.

By the same token, it is important not to fertilize crops with high-nitrogen natural fertilizers when the soil is too cold, for instance very late in the season inside tunnels or the greenhouse. Nitrogen might then accumulate in the form of nitrates, which can lead to toxicity in the vegetables. This can be a problem when growing winter spinach and other greens during periods of low light intensity.

Phosphorus (P)

Phosphorus is required for root development in young vegetable plants and plays a key role in the formation and maturation of fruits and tubers. Like nitrogen, phosphorus is mineralized from organic matter, so all practices that increase biological activity in the soil also increase phosphorus availability to plants. Regular applications of compost and manure should provide the soil with enough phosphorus to meet the needs of vegetables. In fact, the concern with phosphorus in organic farming might be accumulation, not deficiency.

Phosphorus is not very mobile in the soil, and vegetables need less of it than they do of nitrogen. If your soil is in good condition for root development (i.e., not compacted), you should not have to worry about phosphorus. However, when fertilizing heavily with compost and manure, aiming to supply sufficient nitrogen, phosphorus can rapidly accumulate in the soil and pollute the environment via leaching and runoff. This is the main reason for watershed agricultural pollution. And yes, even an organic grower can be responsible for water pollution. Growing leguminous green manures is one solution to this problem as this increases nitrogen in the soil without adding phosphorus.

Potassium (K)

Potassium is the final element in the famous N-P-K equation. It plays a role in allowing root vegetables to keep for long periods and has a positive effect on the size, color, and even taste of fruit vegetables. It also makes plants more vigorous and resistant to diseases, parasites, and adverse weather.

Unlike nitrogen and phosphorus, potassium is not mineralized from organic matter; it is already present in most soils in its mineral form, mostly in the clay fraction. Potassium is very mobile in the soil, which in one way makes it easily available to plants, but in another more prone to leaching. It is one important nutrient that is lost when a compost pile is left bare.

Most vegetable crops require a lot of potassium, but, fortunately, the natural fertility of most soils,* coupled with crop rotation practices and regular additions of compost and manure, will be sufficient to meet the needs of most vegetables. Sandy soils and greenhouse soils, where production is very intensive, are the exception, and in those cases, supplying extra potassium might be necessary. The most common short-term fix for a potassium deficiency is to use a mineral amendment such as potassium sulfate along with an organic fertilizer (manure and compost).

Many books on organic agriculture recommend using mica or basalt to gradually correct a potassium deficiency in the soil. However, a number of experienced organic greenhouse growers, who tried this solution for a number of seasons, told me that these amendments are too slow to act, and that even with large doses, they did not

---

* Soil tests will show if the potassium level is high enough for the crops.

significantly improve the potassium availability in their soil.

Calcium (Ca) and the Other Secondary Nutrients

Calcium, magnesium, and sulfur are often referred to as secondary elements. They play an important role in vegetable growth, and healthy soil generally contains enough of them to meet crop needs.

Having said this, certain sandy or loamy soils may be low in magnesium, which is why some growers use dolomitic limestone (which contains magnesium) in their fields rather than regular limestone. Calcium deficiency causes blossom-end rot in tomatoes and peppers, a common problem in many farms and gardens here in the Northeast. However, blossom-end rot is not a sign of low calcium in the soil but rather a consequence of the plants' inability to assimilate calcium as a result of environmental stress, often irregular watering.

Micronutrients

Micronutrients (also called trace elements) are essential for crop growth, but in tiny amounts. Only the brainier growers can explain exactly what each one does for his crop. In most cases, the existing levels of micronutrients in the soil, good crop rotation, and regular additions of compost should be enough to ward off micronutrient deficiencies in your crops.

However, in some situations, there can be a situation when this is not the case. Examples of this are boron and molybdenum, which are in many areas insufficient to meet the needs of certain crops, especially heavy feeders in the cabbage family. In this case, the two elements can be added directly to the crop via a foliar spray. This will resolve the problem more easily than trying to supplement the soil.

There is a large body of evidence to suggest that micronutrients play an important role in the nutrient quality of vegetables. The importance of remineralizing your soil is something we often hear in advertisements and from convincing salesmen at tradeshows. I don't know how true it is or how much better our crops would be if we bought all these supplements, but we've done fine without them for quite a while now. We do, however, use a compost rich in seaweed, an amendment known to be rich in different microelements.

Good Compost

At Les Jardins de la Grelinette, the organic amendment we use in the greatest quantity is compost, as we believe it is the best ingredient for building and maintaining a healthy soil. Because of its special characteristics, compost cannot be replaced with manure, natural fertilizers (feather meal, bone meal, etc.), or green manure.

Compost is created from the decomposition of carbonaceous organic detritus (straw, leaves, animal bedding, etc.) mixed with nitrogenous material (manure, crop residues, etc.) through a process in which different organisms work on reorganizing this organic matter. When the mix is composed of different ingredients in the right proportions and when decomposition occurs in optimal conditions, the result is a rich and stable

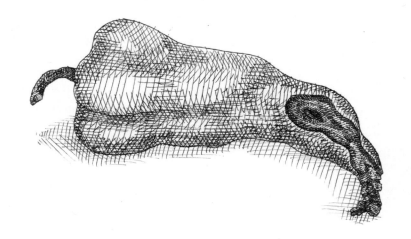

Blossom-end rot is a physiological disease that can lead to major crop losses. It occurs when the weather is warm and there is insufficient consistency of water intake by the plant. This in turn leads to a lack of calcium during the fruiting period. To prevent this disease, we systematically supplement our peppers with calcium during part of their growth cycle.

amendment containing almost all the elements needed to grow vegetables. *Good* compost can supply both the organic matter for soil building and the fertilizer for the crops. It is also packed with soil organisms that activate biological activity in the soil. In other words, it's synonymous with healthy living soil.

I emphasize the word "good" when discussing compost because not all composts are of equal value, mostly because good compost is not so simple to make. Many struggling growers I have met too often fertilize their fields with compost that has either been leached of its nutrients, is only partly decomposed, or—even worse—substituted for by old manure donated by a neighbor eager to get rid of it. In order to grow top-quality crops, a market gardener should understand what goes into making good compost and why raw manure is no substitute. The composting process:

- Stabilizes nitrogen and produces an amendment that releases nutrients gradually over the growing season, even over multiple years. Imagine compost acting like a nitrogen storehouse, a capacity that manures and natural fertilizers do not share.
- Kills potentially pathogenic agents but more importantly the weed seeds that lurk in animal manure, especially that of ruminants. Importing weeds into the garden is a costly mistake that is paid back with extra weeding for many seasons.
- Creates active soil life (fungi, bacteria, earthworms, etc.), which gets transferred into the garden and colonizes it. This added microbial life competes with disease-causing microbes and helps maintain healthy plants.

- Eliminates clumps, resulting in light, homogeneous soil that is easy to shovel and spread over the garden.

Producing quality compost requires a lot of know-how, and, I must confess, it is a science that I have not yet mastered. For this reason, I prefer not to give recommendations on how to make it. One suggestion I might offer, however, especially to beginning market gardeners, is to plan on buying your compost instead of making it yourself. This may sound like sacrilege to some, but in terms of the productivity of the process and the quality of the outcome, it may be the better way of doing things.

A market garden like ours uses so much compost that it becomes quite an undertaking to supply it ourselves. Building a pile is not a problem, we can accumulate organic matter from our crop debris and straw from the neighbor, but turning 40 tons of heavy organic matter regularly, without the help of loading tractor, is a challenge. When considering how occupied we are in the already

"A manure spreader would do the work a lot faster!" This is something I often hear from our interns. Considering that we spread compost on almost 8,000 linear feet of garden bed (every year we use compost on half of our garden—5 plots of 16 × 100-foot beds), using wheelbarrows might indeed feel like a step backward with regards to efficiency. They forget, however, that applying compost with a manure spreader ideally requires not one but two tractors. One is used for hauling the spreader around the garden, and the other is equipped with a loader to make the work go faster. Finding a manure spreader dimensioned for 48-inch bed spacing is also an issue. Considering all the purchases it would require and given the fact that spreading all of our compost takes no more than a week's work, this "solution" is just not right for us. For spreading compost in a market garden, wheelbarrows, shovels, and enthusiastic humans is what I consider to be appropriate technology.

To complete our compost needs, we make our own with crop residues and other available carbonaceous material. When building the pile, we inoculate it with "bokashi" type bacteria, which help to decompose the heap without us having to turn it.

too busy growing season, turning over compost heaps by shoveling them manually is counterproductive. Early on, we came to the conclusion that purchasing commercial compost was the best solution to meet our needs.

We also buy compost to be sure of its quality. Specialized compost companies have the proper equipment and methods needed to intervene at the critical stages of the decomposition process. They will constantly monitor temperature and humidity and turn the pile at the right time. The result of this expertise is a well-structured, homogeneous compost mix which comes with a minimum N-P-K soil analysis. Certain suppliers can also adjust their recipe to accommodate certain

soil types and/or add specific ingredients to their mix on demand. For example, the compost we use contains seaweeds which are high in potassium and micronutrients.

Purchasing a large volume of compost is, of course, an expense, especially with the delivery charges involved. But considering the quality of the product and the time saved, the investment is more than worth it. In our operation, compost-related expenses account for less than 3% of our sales—a negligible amount considering how important this input is to the success of our crops. When delivered, we ask the driver to dump the load in two separate piles located at each end of the garden. Having compost close by, regardless

Whether bought or made, the compost pile should always be covered with a tarp in order to prevent nutrients from leaching away. The pile should also be located on an area where water doesn't accumulate.

of which plot we are spreading it on, is quite a time-saver. We can also have it delivered just before our early springtime seedlings. Incorporating fresh hot compost into a cool spring soil helps to stimulate biological activity and get it going. All in all, buying our compost is so practical I wouldn't want to do otherwise.

When it comes to applying the compost, we use the wheelbarrows to make small piles before spreading it onto our beds with a rake. We then incorporate it into the top two inches of the soil with our rotary power harrow to keep it from drying up.

Relying on Natural Fertilizers—Why?

The poultry manure we use in our gardens is a dried and pelletized fertilizer which, like compost, can be applied just before planting or during the growing season without any risk of bacterial contamination. Its N-P-K value is about 4-4-2 (depending on the supplier), and its nutrients are rapidly available to plants, usually within 30 days after its incorporation into the soil. Unlike the nitrogen in compost, which needs to be worked on by microbes before being released, a good portion of the nitrogen content of poultry manure is in a form readily available to plants, allowing a supply whether the soil is warm or cool.

This is an important feature since, as I mentioned earlier, the fertilizing action of compost is gradual and especially slow in cool spring soils. A combination of both compost and poultry manure ensures that plants will have necessary nitrogen in the early stages of their growth, when they need it the most for a fast development. The manure fertilizer acts like a kickstarter, after which the compost gradually takes over the job of releasing the rest of the required nutrients for optimum growth. This particular combination allows nutrients to be made available as the crops need them and is a crucial aspect of our fertilization program.

From having heard it so many times from interns who've worked at our farm, I know many aspiring vegetable growers tend to express concern about the origins of such poultry manure. It does come from farms where birds are raised intensively in confinement. To a certain degree, it also acts like a fertilizer more than an organic amendment, which seems to emulate conventional agriculture in its principle. These are legitimate concerns that deserve deliberation. In the context of our system, I view poultry manure as a supplementary fertilizer needed for optimum plant growth: it reinforces the compost but does not replace it. It is cheap and easy to apply, and gives good results without being contrary to the natural processes of organic fertilization. For such reasons, I feel this proven solution is called for. Even so, if another product comes around offering the same advantages, I would substitute it without any hesitation. Alfalfa meal is probably one the better alternatives available, but for some reason, it is not readily available in Quebec.

Establishing Crop Rotation

One of the best reasons for growing a wide variety of vegetables is to allow for good crop rotation. Before the advent of monocultures, there was more awareness in the farming community about the two main advantages of crop diversity: allowing soil to keep producing without being drained of its nutrients, while at the same time eliminating a number of diseases and harmful insects that often occur when one species is continuously cropped. When writing farming books before the green revolution (i.e., before about the 1950s), agronomists of that era always advise farmers against using too short a crop rotation. Fortunately, the organic movement continues to emphasize the importance of this time-honored agricultural practice.

For the market garden, crop rotation is basically a process of grouping vegetables by botanical family, and/or by their nutrient requirements, so that they can be grown in alternation at set time intervals. The benefits of this practice are significant but difficult to quantify. Simply put, it improves the cropping system in a number of general ways:

- It disrupts the life cycle of many organisms (insects, diseases, and weeds) which other-wise would be able to take up residence more easily.
- It allows plants with different root systems to penetrate the soil to different depths, thereby improving its structure.
- It reduces depletion of nutrient reserves in the soil by alternating crops with different requirements. We base this alternation on vegetable type: root vegetable, leaf vegetable, or fruit vegetable.
- It helps reduce the weed pressure in the garden by alternating "cleaning crops" with those that have the opposite effect or that require the most advanced weeding techniques (mulching, more frequent hoeing, stale seed-bed technique, etc.)
- It allows for the alternation of heavy feeding crops with light feeding ones. Compost can then be applied one year out of two, for easier management.

In our first years of growing crops commercially, we were not so concerned about crop rotations.

Surprising Advice Regarding Crop Rotation

When starting a market garden, crop rotation is an excellent practice ... to ignore. Alas, the constraints of crop rotation can become such a source of worry that they prevent you from working efficiently. Chances are that, despite all the efforts put into proper planning, the rotation sequence will not be followed. In your first years of growing, it is more than likely you'll decide to add or drop certain crops. It is also almost certain that you will reassess the quantities—therefore the required garden space—of certain crops you want to grow in preference to others. Carefully planning a crop rotation, which will not be respected, amounts to a waste of time. Crop rotation is certainly a necessity over the medium to long term, but you can get away with planting anything anywhere for your first season or two.

We knew about them and understood why they were important, but it wasn't until we attended a seminar where different experienced and successful growers all talked about it—emphasizing how such detail in the planning was beneficial for them in the long run—that we decided to establish a plan ourselves.

Setting up an effective rotation is not so easy, and the implications it entails should not be underestimated. Perhaps more than anywhere else, this aspect of the market garden needs to be carefully considered over time. When ready to do so, I suggest studying different rotations either from books or from talking to organic growers you know, in order to find out the logic behind these rotations. Understanding the underlying principles of why rotations are done is the first stage to being able to develop your own plan. As an example, here is how we designed ours at Les Jardins de la Grelinette.

Crop Rotation at Les Jardins de la Grelinette

The first thing we did when setting up our crop rotation was to consider all of the principles we wanted to follow. Most organic growers plan their rotation around specific field characteristics. Some plots cannot be irrigated, others have soil types favorable to specific crops, some fields are always wet, etc. We took care of all these site characteristics when designing our market garden. We filled the low spots, planned irrigation availability for the whole garden, and took care of drainage, so that we wouldn't have to worry about these anymore. Another common practice is to create

grassland and/or let fields lie fallow for more than one season. Although there are many benefits to this approach, it isn't compatible with an intensive cropping system, and we left that out also.

After looking at many recommendations, we identified these underlying principles which we wanted to follow for our rotation:

- Crops in the Brassicaceae, Liliaceae, and Solanaceae families should not be grown again in the same spot in less than four years. This favorable time interval also applies to the Cucurbitaceae family, but to a lesser degree.
- Heavy feeders are followed by lighter feeders; this makes the best use of compost by reserving it only for plots devoted to heavy feeders.
- Root vegetables are alternated with leaf vegetables.
- Crops that are easy to weed are grown the year before onions, which is one of the most difficult crops to keep weed-free.

Having established these general principles, we then needed to organize them into a pattern for determining how crops would succeed one another and over how many years. As mentioned earlier, the way to do this is to group crops by botanical families and/or nutrient requirements. To help visualize all this, each group can be imagined as a square box that can be moved around. The next step is to play around with different combinations of sequences in order to find a rotation that is consistent with all, or most of, the principles. To make things really simple, we decided at that point (and you will later see the implications of this) that each botanical family (box) would cor-

The book **NOFA Guides Set: Crop Rotation and Cover Cropping** *is a valuable resource for studying different patterns of crop rotation.*

respond to one garden plot. Here is the process step by step.

We knew we wanted to grow a great deal of vegetables of four botanical families: Brassicaceae, Liliaceae, Solanaceae, and Cucurbitaceae, which we considered to be heavy feeders. We also wanted to grow vegetables from three families of light feeders: the legume, Chenopodiaceae, and Apiaceae. Since the light feeders could be combined in any way (without violating any of our principles), we consolidated them together into a fifth "family" to which we added a number of light-feeder crops that belong to botanical families of the heavy feeders. Kale, kohlrabi, arugula, etc. are all short-cycle vegetables that don't tend to carry soilborne diseases, and so we did not see any problem grouping them with the other light feeders. We named this family "Greens and Roots." At this point, we had five families assigned to five different plots.

Plot 1	Plot 2	Plot 3	Plot 4	Plot 5
Solanaceae	Brassicaceae	Liliaceae	Cucurbitaceae	Greens and Roots

Next, we knew we wanted to reduce our compost use by fertilizing only half the garden in a given year. Since the Greens and Roots family is made up of light feeders, it made sense to always precede this family with one of the heavy-feeder families. So, in order to complete the pattern, the rotation plan would have to include four Greens and Roots plots for a total of eight plots. We knew, however, that we wanted to grow lots of garlic. We therefore added a new plot devoted solely to this heavy feeder. This meant we also had to add a fifth plot of light feeders to alternate it with, which brought the total to 10 plots. This sequence allows the spreading of compost to only those plots on which heavy feeders were grown, i.e., one year out of every two, just as we wanted. At this point, the rotation looked like this:

Plot 1	Plot 2	Plot 3	Plot 4	Plot 5	Plot 6	Plot 7	Plot 8	Plot 9	Plot 10
Solanaceae Compost	Greens and Roots	Brassicaceae Compost	Greens and Roots	Liliaceae Compost	Greens and Roots	Cucurbitaceae Compost	Greens and Roots	Garlic Compost	Greens and Roots

We then moved the boxes around making sure that the garlic plot was kept separate from that of the Liliaceae family by four years. Things were looking good at that point, and we knew that from then on, since we had 10 plots in total, the rotation would take 10 years to cycle back to its original starting point.

	Plot 1	Plot 2	Plot 3	Plot 4	Plot 5	Plot 6	Plot 7	Plot 8	Plot 9	Plot 10
Year 1	Solanaceae Compost	Greens and Roots	Brassicaceae Compost	Greens and Roots	Liliaceae Compost	Greens and Roots	Cucurbitaceae Compost	Greens and Roots	Garlic Compost	Greens and Roots
Year 2	Greens and Roots	Solanaceae Compost	Greens and Roots	Brassicaceae Compost	Greens and Roots	Liliaceae Compost	Greens and Roots	Cucurbitaceae Compost	Greens and Roots	Garlic Compost
Year 3	Garlic Compost	Greens and Roots	Solanaceae Compost	Greens and Roots	Brassicaceae Compost	Greens and Roots	Liliaceae Compost	Greens and Roots	Cucurbitaceae Compost	Greens and Roots
Year 4	Greens and Roots	Garlic Compost	Greens and Roots	Solanaceae Compost	And so on... (ten-year rotation)					

Our last step, but not the least, was to make sure that this rotation plan was in line with our production plan. We noticed one thing right away: at any given time, half of the garden had to contain Greens and Roots. This wasn't a problem because we wanted to grow lots of mesclun mix,* a high-demand lucrative crop. Looking into our crop planning, we also realized that we wanted to grow broccoli and cabbage in the spring and fall, but not in the summer. We also wanted to grow a great deal of summer squash, but in two batches: some very early and some later on. To make this possible, we had to compromise a bit and combine these two families together. They would still have a 4-year interval between them. The final crop rotation is as shown on the following page.

Obviously, this rotation plan is tailored to our own production needs, but it provides a good ex-

ample to get started. One of the most important aspects of our rotation is that the way we designed it on paper came to determine how we laid out our whole garden, i.e., 10 equal-sized plots exactly as per our rotation design. This way of doing things has the great advantage of making it very simple to carry out the rotation from year to year, but it does have something of a drawback—our crop planning, deciding what to grow and how much of it, is now determined by the number of beds in each plot.

Since each plot has 16 beds (as explained in Chapter 3), the total number of different vegetables we can grow within the same family is limited to 16. When it comes to planning the Liliaceae crops, for example, we now have to determine that there are going to be 10 beds of onions, 4 of leeks, and 2 beds of green onions. In order to follow our crop rotation, we must adjust production within each family to fit this spatial constraint. We can always make half beds, but we are still limited in overall production.

* To meet different production needs, these plots would have to be replaced by cover crops or other crops that follow the rules established in this rotation.

10-Year Crop Rotation at Les Jardins de la Grelinette

	Plot 1	Plot 2	Plot 3	Plot 4	Plot 5	Plot 6	Plot 7	Plot 8	Plot 9	Plot 10
Year 1	Solanaceae Compost	Greens and Roots	Early Cuc. and Brass. Compost	Greens and Roots	Liliaceae Compost	Greens and Roots	Late Cúc. and Brass. Compost	Greens and Roots	Garlic Compost	Greens and Roots
Year 2	Greens and Roots	Solanaceae Compost	Greens and Roots	Early Cuc. and Brass. Compost	Greens and Roots Compost	Liliaceae Compost	Greens and Roots	Late Cuc. and Brass. Compost	Greens and Roots Compost	Garlic Compost
Year 3	Garlic Compost	Greens and Roots	Solanaceae Compost	Greens and Roots	Early Cuc. and Brass. Compost	Greens and Roots	Liliaceae Compost	Greens and Roots	Late Cuc. and Brass. Compost	Greens and Roots
Year 4	Greens and Roots	Garlic Compost	Greens and Roots	Solanaceae Compost	Greens and Roots	Early Cuc. and Brass. Compost	Greens and Roots	Liliaceae Compost	Greens and Roots	Late Cuc. and Brass. Compost
Year 5	Late Cuc. and Brass. Compost	Greens and Roots	Garlic Compost	Greens and Roots	Solanaceae Compost	Greens and Roots	Early Cuc. and Brass. Compost	Greens and Roots	Liliaceae Compost	Greens and Roots
Year 6	Greens and Roots	Late Cuc. and Brass. Compost	Greens and Roots	Garlic Compost	Greens and Roots	Solanaceae Compost	Greens and Roots	Early Cuc. and Brass. Compost	Greens and Roots	Liliaceae Compost
Year 7	Liliaceae Compost	Greens and Roots	Late Cuc. and Brass. Compost	Greens and Roots	Garlic Compost	Greens and Roots	Solanaceae Compost	Greens and Roots	Early Cuc. and Brass. Compost	Greens and Roots
Year 8	Greens and Roots	Liliaceae Compost	Greens and Roots	Late Cuc. and Brass. Compost	Greens and Roots	Garlic Compost	Greens and Roots	Solanaceae Compost	Greens and Roots	Early Cuc. and Brass. Compostv
Year 9	Early Cuc. and Brass. Compost	Greens and Roots	Liliaceae Compost	Greens and Roots	Late Cuc. and Brass. Compost	Greens and Roots	Garlic Compost	Greens and Roots	Solanaceae Compost	Greens and Roots
Year 10	Greens and Roots	Early Cuc. and Brass. Compost	Greens and Roots	Liliaceae Compost	Greens and Roots	Late Cuc. and Brass. Compost	Greens and Roots	Garlic Compost	Greens and Roots	Solanaceae Compost

Most vegetable growers with whom I've discussed this approach to crop rotation tell me they find it too restrictive, and I have to admit I agree with them. But restriction can be a good thing: it is because we have given ourselves a very specific framework that our rotation is easy to follow. When considering how important the sustainability of our system is, given we want to keep growing intensively on 1½ acres for many years, we believe that such an overarching principle is beneficial.

Green Manure and Cover Crops

Green manures are crops grown not for the purpose of selling, but rather to to add nutrients and organic matter to the soil. They are mainly grasses and legumes, which after being mowed down are then plowed under and incorporated into the soil to boost its fertility. Here are the basic points required to understand green manures.

Many leguminous crops (beans, peas, soy, alfalfa, clover, etc.) have the extraordinary capacity to capture nitrogen from the air and feed it to the soil; this is referred to as "N fixation." When such a crop is turned under, the degradation of plant material allows the nutrient held within the green manure to be released and made available to the succeeding crop. When cereals (oats, rye, wheat, etc.) are mixed in with the legumes, the resulting crop residues provide not just nitrogen but carbonaceous organic matter as well. Green manures can therefore be considered amendments just like compost or animal manure.

The main advantage of green manures is that the primary material used to make the amendment is produced onsite and requires no work other than to seed the crop, shred it, and mix it into the ground. Their disadvantage is that they take up growing space that would otherwise be devoted to vegetable crops. They also take time to grow, from six weeks to a whole season, depending on the chosen plant. An additional two weeks may also be required for the micro-organisms to completely break down the green manure and make the nitrogen available to the plants.

Green manures are generally highly recommended in organic vegetable production. They are an economical and effective way to add nitrogen to the soil, especially when compared with spreading vast quantities of compost and animal manure over large areas. They also fertilize crops with N without adding phosphorus to the soil. However, in the context of a market garden, using green manures as fertilizer is far from ideal. Multiple successions don't allow for much time to grow them, and keeping entire plots fallow does not mesh with the principle of making optimal use of cultivated land. Without the use of a flail mower, it is also difficult to turn them into the soil with a rototiller.

In spite of these drawbacks, we still grow legumes and grasses for a variety of reasons, but we refer to them more as cover crops rather than green manures. Here is how we integrate their beneficial usage into our intensive cropping system.

Adding Supplementary Nitrogen

Although compost is the preferred fertilizer in our gardens, we have not ignored the benefits of fertilizing our crops with leguminous green manures. For us, the trick is to find "holes" in our crop plan that will allow us to precede a crop with a green manure; this is called "catch cropping." Our preferred leguminous cover crops are field peas and common vetch. In both cases, we mix the legume with oats to give it something to climb on as it grows.

The Northeast Cover Crop Handbook is a great resource on different mixes and combinations of cover crops used by organic vegetables growers.

To get the most fertilizing action out of a leguminous green manure, it helps to keep the following two pieces of information in mind.

Firstly, the best time to mix in a green manure is just before it flowers. It is at this stage that the plant has stored its maximum nitrogen and will make the greatest contribution to soil fertility. With the crop still young and tender, it will decompose quickly and easily, in time for the following crop.

Secondly, you need to know that legumes do not fix nitrogen from the air by themselves. The fixing is actually done by bacteria of the genus *Rhizobium*, which form nodules on the roots of the leguminous plants and permit the exchange to take place. However, these bacteria may not necessarily be present in the soil if legumes have not been grown there for some time. Regardless, it is a good idea to inoculate your legume seeds with the proper rhizobia. To ensure that you are getting maximum nitrogen fixation, make sure you inoculate the seeds of each legume species you use. The inoculum is available from seed suppliers as a powder that you mix with your seeds along with some water. Make sure to get some that are in line with organic certification standards.

Adding Organic Matter

As mentioned above, it is good to incorporate a green manure when it is still young and green. Doing so, however, will leave little stable organic matter in the soil after decomposition.

To really add organic matter to the soil by the use of a green manure, very fibrous plants that will resist decomposition must be used. Two good

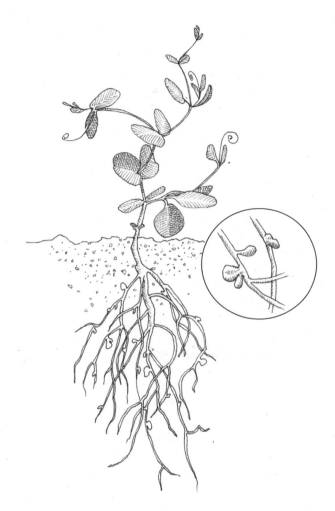

To determine whether to inoculate your green manure seeds, dig up a pea or bean plant (or any other legume) and inspect the roots for pinkish nodules (small roundish bumps). The best time to do this is after the fourth week of growth and before flowering.

examples are fall rye and sorghum-Sudan grass hybrids. When very densely seeded, these crops produce prodigious amounts of biomass while improving the soil structure with their roots. However, because the micro-organisms need to consume a great deal of nitrogen in order to break down the fibrous green manures, they may actually draw nitrogen from the soil. In other words, the normal strategy of turning in large amounts of carbonaceous green manure residue can in fact "steal" the available nitrogen from the soil, leaving the following crop in deficit.

In our gardens, we rely primarily on compost (and lots of it!) for soil building. The only cover crops we grow are those that we seed late in the season to cover the soil for winter. These also add significant amounts of organic matter to our soil.

Protecting the Soil

Leaving the soil bare for several months is a bad idea. Bare soil exposed to strong winds and heavy rains will inevitably degrade in structure and quality. This is especially important in the winter months where these effects are strongest and there is no plant growth to provide protection. In Quebec, snow provides adequate protection during the winter, but when it melts in the spring and saturates the soil with water, runoff can cause serious erosion problems that decrease soil fertility. It is therefore really important to always properly prepare the garden for winter by leaving crop residues in place, covering the soil with a tarp, or by planting a late cover crop that will provide plant cover. Even if this cover crop is one inch long, it's better than nothing. Then, next spring, the soil will have good tilth instead of being hard and compacted on surface.

The ideal strategy, however, is to allow enough time to seed any cereal crop—at least six weeks before the first heavy frost is expected, so that the root system of the plant has sufficient time to develop. If properly established before November, ryegrass will, for example, keep growing even in the cold fall weather and continue early on in the spring. Sometimes, though, it just isn't possible to seed before winter. In this case, we opt for a very early spring cover crop. For example, a pea–oat mix can be seeded as soon as the snow starts to melt (generally early April on our farm) and mowed down after eight weeks, just before the first direct-seeded crops are started.

When planning a cover crop for soil protection, we seed it very densely so that it quickly protects the surface.

Reducing the Spread of Weeds

Large-scale organic vegetable growers grow grasses over a long period to interrupt the weed cycle of their fields. This is not an option for us because we can't afford to lose that much garden space. However, we do take advantage of the weed-suppressing ability of cover crops by planting some between two successive seedings, or in certain beds that we know will be free for a long enough period.

In the middle of the season, our preferred "stopgap" is buckwheat, which forms a plant cover dense enough to suffocate weeds in less than a month. Although this solution is more complicated than installing a tarp and less effective than using the stale seedbed technique (see Chapter 9 on weeding), it is still our preferred option. We like buckwheat because it grows magnificent flowers for the bees and increases biological activity in the soil when turned under. Its tender young tissues are a treat for micro-organisms, which get a boost from it. On many occasions, we've observed how this increase in biological activity favorably affects the following crop.

Buckwheat is not the only cover crop that suppresses weeds. In fact, any species or mix of species can have the same effect, as long as it forms a dense plant cover more quickly than early weeds. The seeding density is crucial in this process, which is why we seed all our cover crops at 5 to 10 times the recommended rate. In our experience, it is much better to spend extra money on seeds than to spend extra time weeding cover crops.

Letting a Plot Rest (full-year fallow)

At some point, we may decide not to grow any vegetables on part of the garden and do so for an entire season. And if this were to happen (don't farmers get sabbaticals?), I would seed a cover crop that would keep growing over a long period, without it being killed when mowed. White clover would then be a good choice: it is hardy, feeds the soil with nitrogen, and is low-growing, thus requiring little maintenance—just a few mowings a year are necessary. However, since clover is a very slow to establish, I would seed a cereal along with it. The first mowing would kill the grain, and by that time, the clover would have had enough time to cover the soil.

If our reason for planting a cover crop is to improve soil quality rather than to let the soil rest, my strategy would be to plant two different cover crops with a fallow period in between. A mix of peas and oats seeded early in the spring would be turned under in June, and then a mix of common vetch and winter wheat would be seeded by mid-August at the latest, to be left as mulch for the winter months. The period between the two crops would be used for two stale seedbeds to eliminate as many dormant seeds as possible.

Establishing Cover Crops and Turning Them Under

We seed cover crops by broadcasting according to the seeding rates listed above. To help ensure good germination, we immediately mix the seed into the soil with a quick and very shallow pass with our rotary tiller or the wheel hoe. Unless especially hot or dry weather is expected (requiring us to deploy a sprinkler line), the moisture in the soil is usually good enough for proper seed germination.

When we are ready to get rid of the cover crop, we first of all shred it using a flail mower. We then have two choices; either leave the cover crop on the surface and cover it with a tarp (and let it decompose with the help of microbes and soil organisms) or till it into the top eight inches of the soil with the rotary tiller. Our minimum tillage practices (see Chapter 5) make us lean towards not using the tiller, but the most common approach is to incorporate the green manure in the lower layers of the soil, providing it with fresh organic matter for soil organisms to decompose.

Cover Crops Used at Les Jardins de la Grelinette

WHITE CLOVER. We prefer white clover to red, which is cheaper but also less vigorous. White clover is slow to establish itself but very hardy. It is also difficult to get rid of once established. We mostly plant it at the edges of our gardens, where it adds nitrogen to the soil, survives the winters, and does not require regular mowing. Since the seeds are so fine, we mix it 50/50 with sand before broadcasting it. This ensures that the seeding is not too dense.
Seeding rate: 2.2 pounds/100 feet of bed (including 50% sand)

OATS AND LEGUMES. Our main catch crop is an oat-pea mix. It can be seeded very early in the spring, as soon as the snow melts. It adds a lot of nitrogen and biomass to the soil and is a great weed smotherer. In the fall, we like to replace the peas with common vetch. Either mix (oats and peas or oats and common vetch) requires eight weeks to grow before being turned under. We aim to seed this cover crop by September 1.
Seeding rate: 3.3 pounds/100 feet of bed, a mix of 60% peas or vetch and 40% oats

FALL RYE. Fall rye is very useful in providing plant cover for beds where late harvests have taken place. This species requires four to six weeks to fully develop and grows even in cold conditions. However, it is very hard to eliminate, and even after shredding, a simple pass with the rotary tiller is generally not enough to kill it. To get our fall rye growing before winter, we aim to seed in the first week of October. The plants come up again in the spring and produce lots of biomass by the end of May.
Seeding rate: 3.3 pounds/100 feet of bed

BUCKWHEAT. Buckwheat is useful for rapidly covering the soil and crowding out weeds. Since it goes to seed eight to ten weeks after being seeded, we take note to mow it before then. Because buckwheat is highly sensitive to frost, late August is our latest seeding date.
Seeding rate: 3.3 pounds/100 feet of bed

For many years, we've incorporated green manures with this consideration in mind. We would then be very careful with the tiller, using its slowest speed so that the tines mix in the amendment without breaking up the soil structure too much. To minimize soil disturbance in this step, the ideal solution would be to use a spade or a spading machine. However, incorporating a green manure using a hand tool on a couple of 100-foot-long beds is too inefficient, and buying a spading machine, considering how rarely we would actually use it, would not be cost-effective.

Currently, our usual way to deal with cover crop residues is to shallowly mix them into the soil with a quick pass of the rotary power harrow or rototiller (to a depth of about three inches) before

A flail mower will shred the plants into small pieces; it is an indispensable tool if you want to use cover crops in the market garden.

covering them with a black tarp. At other times, we don't even work the soil. We simply sprinkle compost over the mulched residue and water it, before leaving it to decay under cover. When we peak under the tarp in the weeks that follow, it's amazing to see all the soil organisms at work breaking down the residues. This biological way of working the soil, however, requires considerable time in bed preparation, which sometimes can be problematic. The tillage option is then, for us, an adequate plan B.

Intercropping with Cover Crops

I use the term "intercrop" to refer to cover crop that is seeded on a bed where a vegetable is already growing, so that when the crop is harvested, it (the cover crop) can occupy the growing area more rapidly. A carrot bed for example would be seeded with drilled clover four weeks before harvest, giving it a head start even before the carrots are taken from the garden. The purpose of this technique is to increase the window of time in which a cover can be used, either before or after

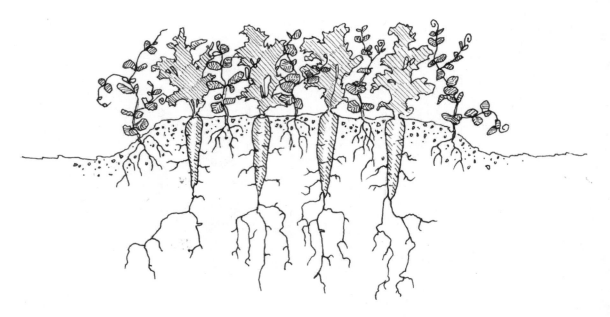

Intercropped green manures are very appealing because they make optimal use of space and time. However, they are not easy to integrate into an already complex crop plan.

the main crop (or both). In theory, this is a highly attractive option for intensive cropping systems like ours.

In practice, however, the technique has not convinced us. On one hand, after trying it a few times, we found that the cover crop did not grow well between vegetable plants, mostly because of the shadow cast by the main crop. Perhaps our timing wasn't good or our species combination wasn't the best, but our results haven't been as satisfying as when seeding a cover crop by broadcasting it on an empty bed. We also struggle with this practice because of the complexity it brings to our already hard-to-follow crop planning schedule. I mention it, however, because I feel it is a promising idea for the market garden.

Integrating Cover Crops in the Fertilization Program

Although we find cover crops beneficial in a number of ways, including them in our fertilization program has not been easy. Because of the time and space constraint required to grow them, a systematic approach to their use becomes inevitable. The way we manage this is by, first of all, grouping together crops planted on the same dates (more or less) in our annual crop planning (see Chapter 13). By doing so, we can grow cover crops on larger areas with multiple beds, as opposed to many areas where individual beds are planted. Ideally, we like to plant entire plots or half-plots at a time. This just makes more sense from the point of view of managing them.

In order to plan for cover crops well in advance, without it becoming too complicated, we've integrated them into our crop rotation. According to this plan, half our vegetable plots are in green manures for a part of the season.

Connecting with Soil Ecology

Over the course of this chapter, I have been trying to convey how taking an active interest in the living aspects of the soil can help us better understand how to use it wisely. I've tried to explain

Rotation Plan Including Cover Crops

	Plot 1	Plot 2	Plot 3	Plot 4	Plot 5	Plot 6	Plot 7	Plot 8	Plot 9	Plot 10
Year 1	Solanaceae Compost	Greens and Roots	Early Brassicaceae and Cucurbitaceae Compost / *Vetch Oats*	*Rye* / Greens and Roots	Liliaceae Compost	Greens and Roots	*Peas Oats* / Cucurbitaceae Compost	*Rye* / Greens and Roots	Garlic Compost	Greens and Roots
Year 2	Greens and Roots	Solanaceae Compost	Greens and Roots	Early Brassicaceae and Cucurbitaceae Compost / *Vetch Oats*	*Rye* / Greens and Roots	Liliaceae Compost	Greens and Roots	*Peas Oats* / Early Brassicaceae and Cucurbitaceae Compost	Greens and Roots	*Rye* / Garlic Compost
Year 3	Garlic Compost / *Rye*	Greens and Roots	Solanaceae Compost	Greens and Roots	Early Brassicaceae and Cucurbitaceae Compost / *Vetch Oats*	*Rye* / Greens and Roots	Liliaceae Compost / *Rye*	Greens and Roots	*Peas Oats* / Early Brassicaceae and Cucurbitaceae Compost	*Rye* / Greens and Roots
Year 4	And so on... (ten-year rotation)									

Half the plots in production are reserved for cover crops over part of the season. In most of the gardens dedicated to Greens and Roots, there is also a window of time in which buckwheat could be planted as cover crop, but we do not plan for this as we often prefer to do stale seedbeds in these plots or cover them with a tarp.

how and why agronomic recommendations are useful and how you need to feed the soil *and* the plant in an educated way. I went over the practices we apply in our garden, trying to demonstrate how much easier it is to follow complex practices once they are integrated together in a systematic fertilization program. Such guidelines can be of assistance to anyone new to market gardening, and I believe the recommendations we have followed may also prove to be excellent for growers elsewhere.

However, to obtain optimal results, it is also important that you develop your own botanical and soil sensitivity. Reading about soil biology and taking the time to reflect on life within the soil allows us to form a solid relationship with the fascinating world under our feet. With educated observations you can learn a lot about how plants and soil interact—and about how you can intervene to make these interactions even healthier.

At Les Jardins de la Grelinette our objective has always been to create a cropping system that strikes a balance between yield, long-term fertility, and efficiency. If our observations are any indication, I believe that we are on the right track. Nevertheless, our fertilization methods are still a work in progress, and many questions remain. For example, a time-honored principle in organic agriculture postulates that if a plant receives perfect nutrition from balanced soil, it will not be vulnerable to diseases or insect pests. This is definitely not the case in our gardens. We've also done nothing yet to integrate different bioactivators into our fertilization program, just as we are only now learning how to inoculate our permanent beds with mycorrhizae to encourage a fungal community in our gardens. To this end, we have been experimenting with ramial chipped wood (RCW).* We also want to learn how to use compost teas and other applications of beneficial bacteria in order to stimulate the life in our soil differently. I have not discussed any of this, because our experimentation is still in progress. But whatever results our experiments yield, we know for sure that there will still be much to learn and discover when it comes to biological solutions. As organic growers, we have the opportunity of practicing applied ecology every day of our lives. This is a great privilege.

* RCW is a technique in which ligneous material is introduced to the soil to increase the presence of beneficial fungi. More information on this practice can be found in the bibliography and appendices.

Ten Tips for Organic Crop Fertilization

- Stimulating biological activity in your soil should be your priority. You must constantly work to encourage soil that is well structured, well aerated, moist, and warm.

- Biological activity is greatest at a pH between 6.2 and 6.8. Lime accordingly, aiming for a pH of 6.5.

- Organic matter is both food and habitat for soil micro-organisms. In the beginning, focus on building your soil and aim to achieve a high level of organic matter as quickly as possible. Later, focus on replacing the nutrients taken up by your crops by adding organic amendments, including compost, animal manure, and perhaps green manures.

- Feed the soil, but also feed the plant. This involves addressing the different crop requirements, the nutritional value of the inputs, and the role of each nutrient—especially nitrogen in the early stages of crop development.

- Soil testing is an important tool. It can detect potential mineral imbalances, tell you the natural fertility of your soil, and help you to determine any fertilizer supplements you may need.

- Since compost is the principal source of fertility, it must be of high quality. If you cannot guarantee that you will be able to make compost efficiently and knowledgeably, it is best to buy it. Make sure it is made from as many sources as possible and amended with a rich source of micronutrients. Compost piles should always be well covered to keep nutrients from leaching.

- A healthy crop rotation will allow you to avoid many crop problems and will ultimately be what makes your system sustainable. Give yourself a few seasons to develop a good knowledge of vegetable production before implementing a complex crop rotation.

- Organic crop fertilization involves integrating a plethora of different agricultural practices. To manage this complexity , work on developing a coherent fertilization program. A systematic approach will, in the end, make your day-to-day operations much easier.

Starting Seeds Indoors

To have a green thumb: to be gifted at caring for plants.

— Common saying

MOST VEGETABLES on our farm begin life as seedlings in a nursery that we set up each year to grow transplants. Given the choice between transplanting a crop and direct seeding it, we always prefer to transplant. The advantages of this method are worth all the effort and expense, especially if intensifying production on a small plot is the goal. The success of our growing operation therefore depends on our ability to start seeds indoors. Failed germination, slow growth, diseases, or any problems at the seeding stage can have disastrous consequences on the production calendar we rely upon for timely harvests. More than at any moment in the season, indoor seeding requires competence and attention to detail.

Having said this, producing transplants is only a small part of what we do in a season, and it is hard to justify investing in the most cutting-edge facilities and equipment just for this. The science of greenhouse horticultural production is very advanced, and mastering the best techniques would be a huge undertaking. In spite of this, we have managed to learn enough about indoor growing to suit our purposes and have developed systems for producing very high-quality transplants in a low-tech setup. For successful market gardening, this balance is important—one needs to develop the skill to produce large quantities of vigorous and healthy transplants without having to become a specialist. Here are the principles we follow at Les Jardins de la Grelinette.

Seeding in Cell Flats

There are many techniques for starting seeds indoors. Most amateur gardeners use open polystyrene flats or individual pots made of coconut fiber. On the commercial side of things, Eliot Coleman has been a long-standing proponent of soil blocking, a technique in which seeds are germinated in blocks of soil created by a press. Having tried this as well as other techniques, we settled for the more common way of growing transplants in cells (also known as plugs), which is an effective and proven technique that I strongly recommend.

Cell flats are plastic containers separated into many compartments in which the roots of the seedlings begin to grow. Flats are placed in

Transplanting is the practice of starting seedlings in one place and setting them out later on in the garden.

containers called trays that add support for moving them around. Most flats are 11 inches wide and 21 inches long, an industry standard that allows for uniformity in tables and other nursery equipment, such as harvest carts for transporting transplants.

Planting cells come in a variety of sizes. It is important to select the size of plant cell for the particular type of plant you are growing. A chart of recommended plant cell sizes can be found at the end of this chapter.

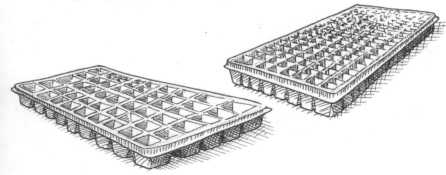

Flats have anywhere from 24 to 200 cells, which determines how big the cell size is. You then choose the correct cell size in accordance with the soil volume required by the roots and with the number of days each crop will spend in the cell. Each plant's root system will develop within its own cell, making it easy to transplant them individually to specific sites or containers (their roots do not become tangled). In our nursery we use flats of 72 and 128 cells, and bigger individual pots of four inches in diameter for seedlings that are potted up.

There are many advantages to working with cell flats: they are easy to handle and fill, drain well after watering, and create clumps of soil that hold together well—which is one of the keys to successful transplanting. They are also durable and reusable, although not entirely indestructible. A few always make their way into the garbage by the end of each season, which is certainly one of their drawbacks. Since flats are used from season to season, you must make sure that they do not become vectors of different plant diseases. In our years of using flats, we have never washed or sterilized them, and have never had any problems. We do, however, always empty ours completely and spread them out to dry in the sun for a few hours at the end of each transplanting session.

The Soil Mix

Working with the proper soil mix is a very important aspect of growing seedlings in cell flats. Since all basic plant needs (air, water, minerals, etc.) must be met with only a small amount of growing medium, the ingredients of the mix must have specific characteristics (drainage, water retention, aeration, fertilization, salinity, pH, etc.). A soil mix is therefore not something that can be improvised, and for this reason, buying a commercial mix is perhaps a better option than making it yourself. When doing so, it is important to choose a top-quality product and make sure it is not treated with synthetic wetting agents. Most certified organic mixes are suitable.

This being said, preparing one's own mix is neither difficult nor complicated. Here is an "all-purpose" recipe that we have used successfully for many seasons. (Our buckets have a volume of 4.2 gallons.)

- 3 buckets peat moss
- 2 buckets perlite
- 2 buckets compost
- 1 bucket garden soil
- 1 cup blood meal*
- ½ cup agricultural limestone

In the United States and Europe, coconut fiber is often presented as a green alternative to peat moss, but I have my doubts that importing precious organic matter from the tropics is more environmentally friendly than using a local resource.

These ingredients are found in most mixes, and you can research their origins and characteristics if you wish. There are certain details to note, however:

- Peat moss is the main component of the mix and should be top-quality. Avoid peat moss that is too coarse or too fine.
- Perlite serves as an aggregate and plays a key role in drainage and aeration. In this recipe, it can be replaced with vermiculite, especially when used in larger cells (flats of 72 cells or fewer).
- Compost must not be leached of its nutrients

* When potting up with this recipe, we double the amount of blood meal.

and must be fully mature (i.e., completely decomposed and not warm to the touch) in order to avoid germination problems. We use the same compost for this as we use in our gardens.
- Garden soil is used to "cut" the compost and to lower the salinity. Use a light soil (not too sandy, not too heavy with clay). I prefer to use soil from our garden rather than sterilized garden soil so that living material is introduced to the mix.
- Blood meal supplies the extra nitrogen required by heavy feeders. Feather meal can be substituted for blood meal in this recipe.

- Agricultural limestone is used to raise the pH of the mix, which has a tendency to be low owing to the natural acidity of peat moss.

The mixing can be done directly in a wheelbarrow. For best results, we mix the limestone into the peat moss first. The rest of the ingredients are then incorporated one by one with a shovel. Working with dry ingredients makes for a more homogeneous mix, but it's also important that in the end the mix be well moistened throughout. We find it best to water it as we go along in the process of mixing all the ingredients.

The mix must be sifted to remove rocks and other large debris to ensure a consistent product. When watering the flats, uniformity in the cells will make an important difference. We sift our mix using a wooden frame with a metal mesh stretched across it. The holes of the mesh are about half an inch square. All and all, making your own mix is a simple process, but it can be tiring. There are, however, some ways to ease the workload. For example, some farmers use cement mixers to stir their soil mix.

Filling Cell Flats

Flats must be filled with the proper technique in order to conserve as much air retention capacity as possible in the soil mix. This makes a big difference with regard to proper root development of the seedling. Our process is as follows.

The first step is to make sure the mix is thoroughly wet. It should be sticky—just before the soil holds together in a ball. If it is not wet enough, we add more water to the mixture, which we stir again with a shovel until it is just right.

The cell flats are then filled to the brim with the sticky mix, and the excess is cleared away with a straight piece of wood or a brush. Once again, it is important that the cells be filled uniformly: those with less soil will dry out more quickly, which makes watering a more difficult job.

The mix is then packed lightly, by lifting the flats about two inches into the air and letting them drop onto a flat surface. Once the cells have been

For a number of years, we have organized a garden plant sale—to great commercial success. Since our production of transplants has increased so much, we have started using a purchased certified organic soil mix. Once production reaches a certain level, this option becomes very attractive.

seeded, we cover the seeds with a fine layer of dry soil mix to prevent them from drying out too quickly. In the end, the flats should be five sixths full to ensure enough space for water retention.

The very last step is to arrange the seeded flats in the nursery. It is important that flats with the same number of cells be grouped together to ensure uniform watering.

The Seedling Room

Certain crops that we want to produce early (e.g., tomatoes, leeks, and onions) must be started as soon as spring arrives to be ready to harvest early in the season. Given how expensive it is to heat a greenhouse during the very cold month of February, it is a good idea to start the seedlings inside your house, in a place that is already heated, where flats can be spread out and watered, and where soil mix can be handled. We refer to this space as a seedling room, and there are many ways to create one. Here are the main factors to consider when planning one.

The main objective of a seedling room is to be able to control growing conditions perfectly. The ideal average temperature for plant growth is 18°C to 23°C (64 to 73°F) in the daytime and 18°C (64°F) at night. The temperature in our seedling room is regulated with baseboard heaters, but any heating system will do, provided that the heat is not blown directly onto the plants. The relative humidity in the room should be 60 to 90%, which is easily maintained with a timer-controlled vaporizer (e.g., this can be set to mist for 10 seconds every 20 minutes). The seedling room should be lined with polyethylene to trap in heat and humid-

To avoid compacting the cell flats when filling them, make sure they are stacked at alternating angles rather than inserted into each other.

ity. A small fan is also helpful to prevent fungal disease resulting from a buildup of stagnant air.

Because the day length in February and March is too short for optimal plant growth, extra lighting is require so that the plants receive 14 to 16 hours of light per day. There are a number of solutions for this, but the simplest and cheapest is to position fluorescent tubes above the flats. In order to provide the plants with the full spectrum of light, you will need both cool white and warm white fluorescent bulbs; the latter produce infrared radiation and are generally designed for bathrooms. To prevent plants from withering, the lights should be adjustable in height and positioned about four inches above the tips of the plants as they grow.

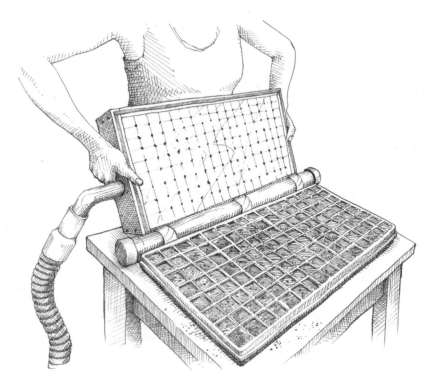

To speed up the procedure when seeding flats, we use a homemade vacuum seeder, also known as a plate seeder. This simple yet brilliant device uses a vacuum to hold seeds onto a flat plate drilled with small holes to match the seed size and pattern of cells in a tray. You then flip over the plate onto a filled cell flat and shut the vacuum off, letting the seeds fall into place.

Most vegetables require a higher temperature to germinate than they do to grow. In order to ensure optimal growth, the seedling room can be equipped with heat mats that plug into a 110-volt outlet. With these mats, the soil temperature is kept at an optimal germination temperature, about 25°C (77°F) day and night. Contrary to popular belief, darkness does not help seeds ger-minate—but soil humidity does, which is why we water frequently and sometimes keep our flats under floating row cover.

The Evolving Plant Nursery

A seedling room is ideal for starting a small amount of indoor production, but once a certain number of seedlings are started, it becomes in-evitably necessary to move into a bigger space. A greenhouse dedicated to the production of trans-plants is therefore essential in any market garden-ing operation. This requires a major investment in infrastructure, for the building as well as for heat-ing and ventilation equipment.

When we planned our nursery at Les Jardins de la Grelinette, the option we found that made most sense was to set it up in a greenhouse that would also be used for growing crops during the summer season, rather than investing in a smaller permanent greenhouse intended exclusively for producing transplants. Our reasoning was that since we needed a nursery only for about 12 weeks out of the year, having multiple uses for one build-ing made more economic sense. We also liked the idea of using some of the heat provided for the nursery for other early direct-seeded crops as well. Our evolving plant nursery addresses both these objectives.

At the end of the winter, we set up the nurs-ery in part of our large tomato greenhouse, which we section off with a piece of polyethylene film. The partition is clipped to the hoops of the green-house, making it easy to move. This allows us to increase the heated space gradually as our trans-plant production progresses. Since only part of the greenhouse is being used as a nursery, we seed the remainder of the space with early vegetables that we plan to harvest for our first market. All in all, the nursery takes half a day at most to set up. It is simply a matter of laying a geotextile ground cover (to prevent weeds from taking root) and installing movable tables.

As the spring wears on, the outside temperatures get warmer, and some of our inside seedlings are transplanted out into the tunnels and under floating row cover in the gardens. We take this opportunity to reorganize the nursery, reserving a section for the tomatoes that are ready to continue growing in the ground. The polyethylene partition is removed, which allows the greenhouse to warm the tomatoes, the transplants, and the early crops. Once the danger of frost is behind us, we move all of the transplants outside. In our climate, this usually happens around the end of May or the start of June. At that time, we also harvest the greenhouse crops (usually beets and carrots) for the first market, and the rest of the tomato plants are quickly planted over the entire greenhouse space.

All of this involves a lot of handling and planning, but it allows us to use the heated space optimally. Considering how expensive fuel can be, this game of musical chairs is well worth the effort.

Heating and Ventilation of the Nursery

No matter how a nursery is organized, it needs proper heating and adequate ventilation to produce good transplants.

One classic mistake beginning growers often make is to set the greenhouse thermostat lower

In our house, we use a hallway with south-facing windows as our seedling room. This space is next to the family kitchen, so it is easy to keep an eye on everything.

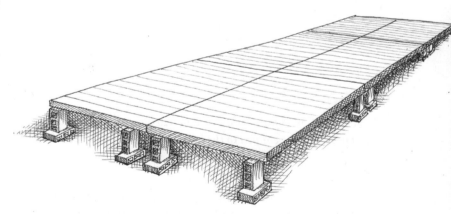

The wooden seedling tables we use are placed on concrete blocks. These are easy to build and move. To make the best use of the available space in our greenhouse, we have some tables that measure 4' × 8' and others that measure 2' × 8'.

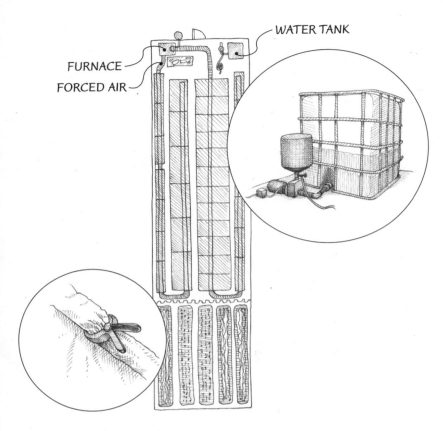

FURNACE

FORCED AIR

WATER TANK

The large tomato greenhouse is divided in two by a piece of polyethylene clipped to the hoops.

Purchasing a furnace for your greenhouse is one place where you don't want to be cheap. Get a new one (or refurbished) and make sure it has enough BTU to rapidly heat up the space.

than the optimal temperature for plant growth (18°C/64°F at night) in an effort to save money. This is understandable considering how often the furnace will run on through the night—but in the end, this amounts to cutting corners. Plants kept at a lower temperature will grow more slowly, which delays production. Given how short our growing season is, we have to make transplants grow as quickly as possible. When it comes to saving on heating costs, it is better to improve equipment (e.g., better greenhouse insulation, more efficient furnace, thermal screens, etc.) rather than adjust the temperature. Especially important is making sure that the greenhouse is shut up tightly at the end of the day to keep out the cold wind at night.

Regarding which heating system is better—oil, propane, or gas—they are roughly the same in terms of heating and equipment cost. In my opinion, none of them is environmentally friendlier than the other. Heating with a wood furnace is also an option, but one that I do not recommend. Waking up a couple of times per night to fill the stove is a hardship, and wood furnaces aren't as easy to keep at a set temperature. With heating systems, the most important thing is to acquire a furnace that is in excellent condition, powerful, and above all, reliable. It's also important to make sure that the furnace is the appropriate size. A small furnace that takes forever to heat the space will actually burn more fuel than a large one that brings the greenhouse to the correct temperature quickly. It's also worthwhile to investigate the reliability and speed of the services offered by different fuel suppliers. Some have special rates for farmers, so it is always a good idea to ask. Also, to avoid a fuel shortage, *always* ensure that your tank has plenty of fuel on nights when you expect frost or very cold weather.

To ensure uniform heat distribution, perforated polyethylene pipes may be used; these are attached to the furnace and installed under tables in order to heat the seedling flats first. The perforations must be calibrated according to the size of the pipes and their spacing inside the greenhouse. Most greenhouse equipment suppliers offer this service.

To cool the nursery on sunny days, different options are possible. We decided to go with natural ventilation in the form of roll-up side walls. While there are a number of forced air fans available that can ventilate more precisely, we appreciate the passive quality of this system that simply uses wind to clear the air. To keep the cold air from touching the transplants too directly, we installed "skirts" along the greenhouse and place our seedling tables below the level of the side openings. This is a very common feature that greenhouse manufacturers can help you set up.

When it comes to regulating humidity, we have gotten in the habit of rolling up the sides for a few minutes very early in the morning to allow any humidity produced through condensation during the night to escape. I also recommend doing so *while* the furnace is heating the air. This procedure prevents excessive humidity from stagnating in the nursery and works wonders. In all of our years growing, we have had no major fungal disease in our nursery to date.

Finally, one of the most essential tools in any nursery is a thermometer with an alarm that allows you to set maximum and minimum temperatures. In the event of a furnace breakdown, power failure or fuel shortage, our alarm alerts us that our seedlings are in danger. This is particularly important on frosty nights, when a few hours without heating can be fatal to plants. We've had such circumstances happen more than once, and fortunately, we had a backup plan. If the furnace breaks down, we have an auxiliary furnace in the greenhouse that we maintain regularly. This second furnace is much less powerful (and less expensive!) than the main one, but it keeps the greenhouse just warm enough until the main one can be repaired. After several seasons of denial, we also reluctantly bought an emergency generator for use if the power were to go out for several days. This was a large expense and will probably never be used, but we would stand to lose if we didn't have it available as a backup. As my grandfather used to say, "Sometimes you need to wear suspenders with your belt."

The alarm on our thermometer also warns us if the greenhouse is at risk of overheating during the daytime. Forgetting to roll up the sides when the sun comes out can result in a greenhouse full of dead seedlings in less than two hours! Such a catastrophe would be disastrous for the whole length of the growing season, and therefore this is not a wise place for gambling. An alarm thermometer is absolutely indispensable in the nursery.

How to Water Seedlings

Water management is essential for producing successful seedlings. Like an overheated greenhouse, a water shortage can quickly kill plants—but, on the other hand, too much moisture in the soil mix can lead to fungal diseases. The question of when

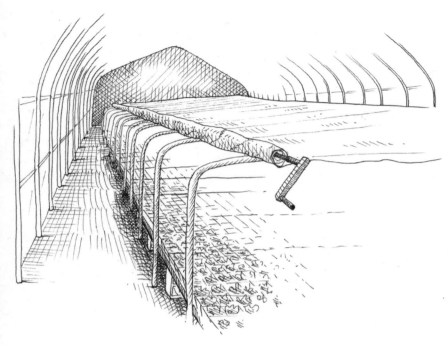

It is a good idea to cover seedlings with a thick floating row cover or even a polyethylene film at night. There are different designs that make it easier to cover and uncover the plants. Thermal screening is an inexpensive way to reduce heating costs.

- It's important to water in accordance with the location of the trays in the greenhouse. Flats generally dry out faster at the table edges, on the south side of the greenhouse, and closer to the furnace.
- Watering is a task that must always be done twice: once to moisten the soil via capillary action and once to water deep down.
- When watering, it's important to consider the outside temperature: sunny weather calls for heavy watering, while cloudy weather calls for light watering—or none at all. Devastating diseases such as "damping off" arise in soil mix that stays soggy for too long. Also, plant leaves that remain wet for too long can act as an entry point for various fungal diseases.

Managing all these factors requires special sensitivity that one gradually develops through careful and continuous observation of the growing medium and the appearance of the seedlings. For this reason, it's a good practice to have only one person do watering inside of the nursery. Having the same person in charge also ensures that this essential task is never forgotten. This person may delegate on occasion, but never for too long. This is a policy I highly recommend.

It is also important to make sure the water used is not too cold, as this will slow the growth of the seedlings. Our solution is to have a large water tank (265 gallons will do), which we fill up every other day. The heat of the greenhouse warms up the water, which would otherwise come directly from our well. To heighten this effect, we painted the reservoir black. This is also a good way of pre-

and how much to water is a complicated one, involving the consideration of many factors:

- Watering must be done in a uniform fashion. Every tray of the same cell size should get the same amount of water, so that none of them dry up faster than others. Obviously, pots and trays with bigger cells will require more water. It's therefore important to organize the layout of the flats accordingly on the tables, i.e., flats of 72 cells together on one set of tables and those of 128 cells on another.

venting algae formation in the tank. The reservoir is linked by pipe to a pool pump that is equipped with a pressure tank. This allows us to water as much as we need without always having to switch the pump on and off.

Potting up

Potting up is a technique that consists of transferring seedlings from small cells into bigger ones. This practice gives plants that spend a long time in cells (tomatoes, peppers, cucumbers, and eggplants in our case) the benefit of extra root space and rich new soil mix to further their development.

The procedure is simple yet delicate. The seedlings are fragile and can experience stress if their roots are damaged. When potting up, we make sure to hold each plant by its stem and extract it from its cell by pinching the cell bottom and gently tugging the plant. If plants are potted up at

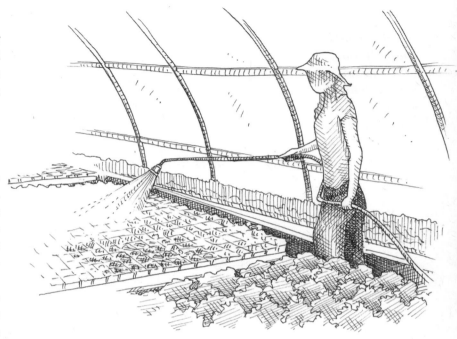

Watering seedlings is a fine art: finding the balance between watering too much and not enough requires attention to detail.

Timing Your Production Right

The crop plan, which we use to plan our yearly production, is largely based on the scheduling of indoor seedings. In order for plants to take off quickly after transplanting, they must not be kept in their cells for too long. Every plant has a set period within which it will do well in cell flats. We use this information, in conjunction with the desired transplant dates, to schedule our seeding dates.

It is also important to precisely determine not just when your transplants will be ready but how many of them you will need. There is no point in growing too many, considering that seedlings taking up greenhouse space are costly to produce; yet you don't want to be short on plants when the transplanting date rolls around. We use a chart like the one presented on page 94 to plan the number of trays we need to start, and to properly estimate timing.

the correct stage, the roots will generally occupy enough space in the cell so that the clump holds together well. To ensure that we grow only the most vigorous transplants, we never pot up weak or sick seedlings—these go into the compost along with their used soil mix.

Transplanting into the Gardens

Transplanting is an exciting time. After months of nurturing the seedlings, they are now ready to fill up the garden, which then quickly takes form. However, most of the transplanting also happens during spring rush, when everything needs to be done all at once. Therefore, we try to organize this operation as efficiently as possible.

The first thing we do is prepare the plants for the shock they will have when moved outside. Since they have spent their whole lives up to this point in an ideal, regulated growing environment, they are not yet used to wind, rain, and outside temperature variations. One week before the transplanting date, we put the seedlings on tables just outside the greenhouse and cover them at night with floating row cover. If frost or very

cold temperatures are expected, we bring them back inside the greenhouse. The purpose for this is to get the plants accustomed to the elements gradually.

During the hardening off period, we prepare the beds and make sure they are ready to receive the transplants: soil is amended, beds are broad-forked, and plastic mulch and drip irrigation are installed where necessary. We watch the weather forecast very closely at this time in order to seize the most opportune moment for transplanting. We transplant in the morning in cloudy weather and in the late afternoon in sunny weather—but if it is too hot or if the seedlings are transpiring excessively, we postpone the operation altogether.

Once the beds are fully prepared, we make sure to give all transplants bound for the garden a thorough watering. This step is *very* important, as plants take best when there is already a good amount of moisture in the potting mix: the soil in the garden is relatively dry and tends to draw moisture from the transplanted clumps. Therefore, each flat is watered multiple times in order to saturate the clumps with water. They are now ready to be brought out to the gardens in a harvest cart.

Because of our crop plan (see Chapter 13), all of our transplants are to be set out in a specific location in the garden. To make sure there is no mistake about where everything goes, we make notes resembling the ones shown in the table on page 94. This practice also helps us make sure we are not carrying more trays than needed, or that we bring enough. Like I said, we try to be as efficient as possible because we don't have time to spare. The devil is in the details.

Transplanting, if not done properly, can set back the growth of the plant for up to 2 weeks. To prevent this, make sure your transplants are well hardened off before setting them out. This being said, hardening off does require a lot of handling. We actually skip this step for plants that are covered with row cover as soon as they are put in the garden. The growing conditions under the cover are quite similar to those inside the greenhouse, and the crops take fairly well in this environment.

The way we set out our transplants is fairly simple. Working in pairs (or as a trio, depending on the crop type), we work side by side extracting the plants from the cell flats before planting in the ground. To make sure the transplants are spaced correctly, we plant the seedlings in rows that have been previously traced in the soil with a rake (the same raking technique is used to mark the rows for direct-seeded crops). The fastest person of the two (usually me!) will also space the plant on the row using a ruler as a measure.

When putting transplants into the ground, there are two things we watch out for. First, it's important to avoid leaving air pockets between the clump and the hole. The transplants need to be firmly rooted by pressing them lightly into the soil. It's also essential that the clump be buried completely, as it will dry more quickly if left poking out of the soil. By the end of the process, the surface of each clump should be level with the soil surface. These are the instructions we give to everyone who helps us transplant.

For the next few days after transplanting, we make sure the ground stays moist. The plant roots cannot go without water at this stage, or else the crop will be fragile throughout its growth. If the weather forecast calls for sunny or partly sunny conditions, we install an irrigation line to keep the plot watered. Given that crops kept under row cover are particularly susceptible to wilting in excess heat, we take no chances: if it is sunny for the first few days after transplanting, we uncover these crops. This also involves a lot of work, but at this final stage of transplanting, every precaution protects our investment.

At the current time, we are experimenting with a new marking roller that allows us to mark lengthways and crossways at the same time. Clever in its design, this roller can quickly adjust to different spacing. See the tool appendix.

Plot 7, Solanaceae: 16 beds

Bed 1 – Nadia Eggplant: 5 flats
Bed 2 – Beatrice Eggplant: 5 flats
Bed 3 – Hungarian Hot Wax Pepper: 6 flats
Bed 4 – Ace Pepper: 10 flats
Etc.

Transplant Table

Vegetable	Cell flat	Number of flats per bed*	Number of days in cell	Intensive spacing in 30-inch bed	
Asian greens	72	4	21	3 rows	Planted every 12"
Basil	128	3	25	3 rows	Planted every 12"
Beet	128	11	25	3 rows	Planted every 3½"
Broccoli	72	3	30	2 rows	Planted every 18"
Brussels sprout	72	3	30	2 rows	Planted every 18"
Cauliflower	72	3	30	2 rows	Planted every 18"
Celery	72	10	60	3 rows	Planted every 6"
Celery root	72	5	60	3 rows	Planted every 12"
Chard and kale	72	5	30	3 rows	Planted every 12"
Chinese cabbage	72	3	30	2 rows	Planted every 18"
Corn	128	4	15	2 rows	Planted every 6"
Cucumber	72	1.5	15	1 row	Planted every 18"
Eggplant	4" x 4"	85	50	1 row	Planted every 18"
Fennel	72	7	30	2 rows	Planted every 6"
Green onion	500/tray**	13	45	5 rows	Planted every 6" in clumps of 5
Ground cherry	72	1	40	1 row	Planted every 24"
Kohlrabi	72	9	30	3 rows	Planted every 7"
Leek	300/tray **	2	65	3 rows	Planted every 6"
Lettuce	128	3	30	3 rows	Planted every 12"
Melon	72	2	15	1 row	Planted every 18"
Onion	500/tray **	3	50	3 rows	Planted every 10"
Parsley	128	8	40	4 rows	Planted every 6"
Pepper	4" x 4"	170	60	1 row	Planted every 9"
Rutabaga	72	7	30	2 rows	Planted every 6"
Spinach	128	11	21	4 rows	Planted every 6"
Summer cabbage	72	3	30	2 rows	Planted every 18"
Summer squash	72	1	15	1 row	Planted every 24"
Tomato	6" x 6"	170	60	1 row	Planted every 9"

* The number of flats per bed is calculated based on a 100-foot bed length and includes a seeding density 30% higher than necessary. This safety factor is important as it allows us to compensate for losses at the time of planting, or for poor germination in the nursery.

** Crops in the Liliaceae family are broadcast in open trays (without any cells).

Direct Seeding

He who sows densely harvests scantily;
he who sows lightly harvests plenty.

— French proverb

ANY DISCUSSION OF DIRECT SEEDING must begin with the acknowledgement that for the purposes of market gardening it's generally more efficient to transplant seedlings than to direct seed. Transplanting ensures perfect density, allows the crop a considerable head start over the weeds, and greatly reduces the amount of time spent weeding. It's also much easier to ensure optimal germination when operating in a controlled environment. However, there are some crops that just don't transplant well and need to be direct seeded.

Despite the drawbacks associated with direct seeding, it is a faster, easier, and less expensive process than growing seedlings indoors. In this chapter, I will discuss the many different tools and techniques that can be used to simplify the practice of direct seeding. To begin with, two important things must be understood.

Firstly, it's important to ensure a good germination rate for high-yield production. Starting with quality seeds should be a given. Nothing good can come out of seeds with a poor germination rate, so I highly recommend acquiring them from professional seed producers with proven track records. Properly storing the seed stock throughout the season is also very important. Seeds should always be stored in an airtight container and placed in a cool and dry environment. To ensure that the seeds we rely upon are always in prime condition, we try to avoid using seed stock for previous years. We manage to do so by calculating our needs as precisely as possible when placing our annual seed order.

In the garden, germination rates will also be influenced by climatic conditions, which are notoriously difficult to predict. Since uniform growth depends on soil moisture and temperature, our job is to control these parameters regardless of what happens with the weather. This is the main reason why having a reliable irrigation system is essential for the market garden. For fast and optimal germination, the soil should *always* be kept moist right up until the seedlings emerge. In cool weather, it's recommended to cover the beds with floating row covers, which helps keep the soil much warmer.

Secondly, it's important that the crops be spaced with precision for optimal use of the bed space. One easy way of achieving this is to broadcast or heavily seed, and thin the rows to the desired density later. While this can be effective, it's not efficient: thinning a 100-foot row of carrots is laborious and can easily take two people over half a day. For the sake of productivity, it's preferable to seed precisely and keep thinning to a minimum. This is where precision seeders, which can drop seeds at the desired spacing, come in very handy.

Precision Seeders

Manual seeders have been around for a long time, and there are numbers of different models available on the market. Finding the best seeder for the job can be a challenge, as each crop has its own distinctive seed with a specific shape, width, and germination rate. I have tried many of the hand-push seeders currently on the market and found that all have both interesting features as well as drawbacks. Apart from precision, the characteristics that I look for are ease of calibration (i.e., not spending too much time adjusting it for different seed sizes and shapes), ease of use, and price. The ones that we use at Les Jardins de la Grelinette are presented below. You will find a list of the suppliers that sell these seeders in the tools and suppliers appendix.

The **EarthWay Precision Garden Seeder**, also sold under the name Semtout in Europe, is a push-type seeder which drills the seeds into a furrow created by an adjustable opening shoe located at the bottom of the seed hopper. It's driven by a belt attached to the front wheel, which turns a disc called the seed plate. When pushed, the seed plate separates and scoops up the seed before dropping it into the ground. The seeds are then covered by a dragging chain, and the soil is pressed down by the rear wheel. A handy built-in adjustable marker helps you achieve parallel rows when seeding.

EarthWay offers 12 seed plates designed for different seed sizes and in-row spacings (we use only the original 6), which can be changed in 30 seconds. Such easy calibration makes this tool a simple and efficient device to use. The EarthWay is really good for beans, peas, radishes, spinach, and beets,* but is less effective for small seeds, which often get caught in the mechanism. While it may not be the most effective unit on the market, its advantages outweigh the disadvantages, and it is the most affordable on the market. We have never replaced ours since we first purchased it many years ago, and it's a tool I continue to recommend.

* Seeding beets with the EarthWay requires heavy thinning, which is one reason why we prefer to transplant this crop.

The Glaser and EarthWay seeders are very different but complement each other well. They don't provide perfect precision spacing, beets and carrots might not grow to uniform size, but this is not a problem. Our clients don't expect such "perfect" produce.

Like most Swiss tools, the **Glaser Seeder** is simple and well-designed. Rather than operating by a belt, this pinpoint seeder utilizes a ground-driven rotating shaft indented with round holes to carry the seed out of the hoppers and into a furrow. The shaft can quickly be adjusted to three hole sizes (small, medium, or large), which accommodate different seed sizes. A small brush that removes excess seeds provides further adjustment. The user can also adjust the sowing depth by changing the angle of the handle when pulling the seeder along.

Since the Glaser Seeder is very effective for small seeds, particularly round ones, we find that it nicely complements the EarthWay seeder. Its downside, however, is that it only functions properly on finely prepared soil—without rocks, clods, or chunks of un-decomposed organic matter. If any debris is present on the soil, the wheels get jammed and the seeds won't drop. The soil surface also needs to be firmly pressed to facilitate traction. The rotary harrow we use for bed preparation creates the perfect condition for the Glaser Seeder, but a seedbed roller can do the job just as well. Because the Glaser Seeder doesn't have a furrow closing mechanism, we bury the seeds with the back of our bed preparation rake when the seeding is done. Overall, the Glaser Seeder works well, but it requires a certain mastery and skill on the part of the user. I would recommend trying it out first on a small area to get a feel for how it works before using it on a large area.

The **Six-Row Seeder** is designed for intensive seeding and was developed with salad mix and

Depth of planting affects both germination and emergence rate. A good rule of thumb is that seeds should be planted at a depth equal to their thickness, but it's always good to refer to the information provided by your seed supplier.

The Six-Row Seeder is good for seeding crops densely, which is useful in an intensive cropping system. We use it for mesclun mix, baby spinach, carrots, and early radish—when seeding these crops in our hoophouses.

baby greens production in mind. It's the most sophisticated of the hand seeders we use. The mechanism is similar to that of the Glaser, except that the seed shaft is rotated by a pulley attached to two rollers, which act as wheels. These rollers greatly improve the traction of the seeder and tramp down the seeded surface. The tool also has three drive ratios that allow for different in-row spacing densities. Depth is easily adjusted by raising or lowering the front loader.

The hoppers of the seeder are spaced 2¼ inches apart, leaving no room for hoeing between the rows, or for any weeding at all for that matter. The idea behind this design is that after two passes (i.e., up and down the bed) over a width of 30 inches, the seeder will have dispersed seeds so densely that the crop will cover the entire bed, effectively shading out potential weeds. On the one hand, this allows for very efficient use of space, but on the other hand, it requires weed-free soil—an ideal not so easy to achieve. There are different strategies to accomplish this which I will discuss in the next chapter. Like the Glaser, the Six-Row Seeder works best for small-seeded crops (under beet size) and also requires a clean, firm surface in order to function properly.

Seedbed Preparation

Any successful direct seeding begins with well-prepared soil. Hand seeders run most efficiently when the soil is cleared of debris, levelled, and raked to ensure the best possible seed-to-soil contact. The soil surface also needs to be fairly dry in order to avoid clogging up the seeder. If the soil sticks to the surface of the seeder's wheels, it's generally best to wait until the soil is dry.

Seeding in a straight line with rows parallel to one another is also very important as it allows for safely hoeing between rows before the seedlings come up. (This is especially important for weed control in slow germinating crops like carrots.) Before seeding, we mark our rows, and once the bed is seeded, we run the back of the rake over the soil to cover the seeds with a little soil, so that they don't dry out in the sun. We then install a water line and irrigate when necessary. Our water lines are designed to irrigate two, four, or eight beds at a time, so we try to do direct seeding on multiple beds on the same day. For this reason, our crop calendar and garden plan are organized around the seeding dates for direct-seeded crops.

Record Keeping

Nothing is more frustrating than to sow a crop only to find out a month later that the density is

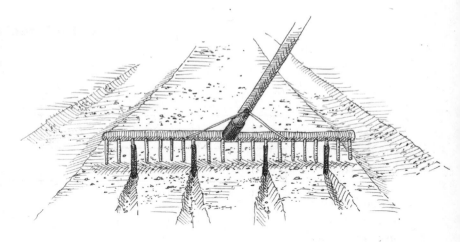

When cultivating with a hoe, it's much easier to follow a straight line than a wandering one. This is why we mark our rows. We do so by fitting a few short lengths of interchangeable plastic tubes over the tines of our bed preparation rake. Our row spacing calculation is based on the width of the hoes we use and the optimal spacing for each crop.

lower than expected. The best way to avoid this scenario is to weigh the seed packets before and after every seeding (the difference being the total weight of seeds sown in the ground) and compare this number to an optimal target density. If for some reason the seeding density is too low, we know to pass the seeder a second time. This simple verification makes it easy to detect a production problem before the crop germinates. If unsure, it's better to err on the side of seeding too thickly than too thinly, even though this might require later thinning of the crop. Systematic record keeping is highly recommended when it comes to seeding, as this might provide explanation in the event of crop failure. The outline below provides examples of how we keep records on our farm.

Notes for Direct-Seeded Crops

Radish: five rows (6" apart) seeded every 1¼ inches, using the EarthWay Seeder radish plate at a depth of ½ inch (density rate between 3 and 4 ounces per 100' bed). Cover with an anti-insect net.

Record keeping has been especially important in determining optimal spacing and in selecting and calibrating our seeders. You can calculate optimal seeding rates for direct-seeded crops by recording the weight of the seeds put in the ground and taking note of the crop spacing later on. Standardised bed lengths simplify this practice.

Plot No.	Number of beds	Cultivar and supplier	Seeding date (2013)	Amount of seeds	Harvest date	Number of days in garden	Yield per bed
2	1	Raxe (William Dam)	May 1	3.8 oz.	June 2 June 9	50	308 bunches
2	1	Pink Beauty (Johnny's)	May 8	3.5 oz.	June 16 June 23	45	285 bunches
6	1	French Breakfast* (Johnny's)	May 25	2.5 oz.	June 27 July 5	48	235 bunches

* French Breakfast seeded in rain; not great.

Direct Seeding Table for Intensive Spacing

Vegetable	Spacing		Seeder	Calibration
Arugula	5 rows	Seeded every 1¼"	Glaser	Large hole: ½" deep
Baby spinach	6 rows	Seeded every 2"	Six-Row	D* (hole size)—L (brush setting)— 2½" (drop distance; middle pulleys)
Bean	2 rows	Seeded every 4"	EarthWay	Bean plate: 1¼" deep
Beet	3 rows	Seeded every 1¼"; thinned to 2"	EarthWay	Beet plate: ½" deep
Carrot	5 rows	Seeded every 1¼"	Glaser	Large hole: ½" deep
Cilantro	5 rows	Seeded every 2"	EarthWay	Beet plate: ½" deep
Dill	5 rows	Seeded every 1¼"	Glaser	Large hole: ½" deep
Mesclun mix	12 rows	Seeded every 1¼"	Six-Row	C* (hole size)—L (brush setting)— 2½" (drop distance; middle pulleys)
Peas	1 row	Seeded every ½"; two passes	EarthWay	Early June plate: 1¼" deep
Radish	5 rows	Seeded every 1¼"	EarthWay	Radish plate: ½" deep
Turnip	5 rows	Seeded every 1¼"	Glaser	Medium hole: ½" deep

*The Six-Row Seeder comes with an instruction manual in which calibration codes are found. The letters presented here refer to these codes.

Weed Management

Too many growers consider hoeing to be a treatment for weeds, and thus they start too late. Hoeing should be understood as a means of prevention. In other words: Don't weed, cultivate... Large weeds are competition for both the crops and the grower.

— Eliot Coleman, *The New Organic Grower*, 1989

AS THE LABORIOUS TASK of transplanting the seedlings into the garden begins to wind down, another job raises its head: weed control. Anyone who has grown a backyard garden knows all too well that vegetables can quickly disappear in a jungle of weeds. So how do you keep a one-acre garden weed-free? And can it be done effectively using hand tools?

Above all, it is important to acknowledge that weeds compete with vegetables for water, nutrients, and growing space. Despite the views advanced in some "natural" gardening circles, the idea that it's possible to grow beautiful vegetables in harmony with weeds is simply untrue; just as seeking a cropping system that would not require weeding in some form or another is unrealistic. So-called natural organic herbicides claiming to control weeds may do so in the short term, but they destroy the long-term biological health of the soil. For weed management practices to be both ecological and sustainable, a market gardener should rather look into careful planning for weed

prevention and follow with effective and efficient weed control strategies. Dealing with weeds the organic way also takes persistence, the right tools, and innovative techniques.

At Les Jardins de la Grelinette, we tightly space our crop to maximize the yields but also to reduce weed growth. Closely spaced plants growing together quickly form a leafy canopy that shades out weeds. This benefit alone is perhaps the best reason for adopting an intensive approach to growing vegetables. We also work with weed-free compost and have adopted tillage tools that don't invert the soil. The fact that we transplant as many crops as possible also contributes to keeping our weeds in check. All of these preventive measures help us fend off invasion of garden weeds, yet they are not enough to keep them at bay indefinitely. We do always have to deal with ground popping weeds attempting to take over our crops.

Our primary weed control strategy is pretty simple: hoe the gardens as often as possible and never let weeds in any plot go to seed. This is easier diminishes at the same time that weed pressure is increasing. But diligence pays off, so the effort is worth it. While weeds still do establish themselves in our gardens, they are noticeably less persistent than in the past. We won't be declaring a permanent victory any time soon, but each season we can spend less time weeding and more time concentrating on other aspects of the garden. Weeding has even become an enjoyable task.

said then done, but focusing on efficiency everywhere else in our operation creates enough time to see this job through. We have also adopted different weeding techniques, here discussed, which make this objective manageable. All in all, keeping our whole garden weed-free is no easy task, of course—especially once the markets begin, and the time available for garden maintenance

Cultivating with Hoes

The bigger the weeds get, the more difficult they are to control. Therefore, the most effective way to deal with weeds is to get to them before they get established, at the stage when lightly disturbing the soil is enough to kill them. For the market garden, the best tool for this job is the hoe.

There are many different kinds of hoes and different names for similar tools. Our favorite hoes are long-handled stirrup hoes with a swiveling, double-sided cutting blade. These oscillating hoes (the ones we prefer are Swiss made) cut through weeds just below the soil surface, both on the push-and-pull motion, so hoeing is very fast, efficient, and ergonomic—which over the long run prevents stiffness and bodily wear and tear. We use the narrower stirrup hoes (3¼ inches) for crops planted in four or five rows per bed and the wider ones (5 inches) for crops grown in two or three rows. We also use a wheel hoe with a very wide blade (12 inches) to hoe crops grown in just one row and to weed the pathways. In addition to stirrup hoes, we also use the collinear hoe developed by Eliot Coleman. This tool is useful for weeding mature crops because its blade can reach

Letting weeds go to seed increases weed pressure for the following year. Galinsoga is particularly difficult to get rid of once it establishes in your garden: it produces about 10,000 seeds per plant! Failure to weed even once can lead to invasions that last many seasons.

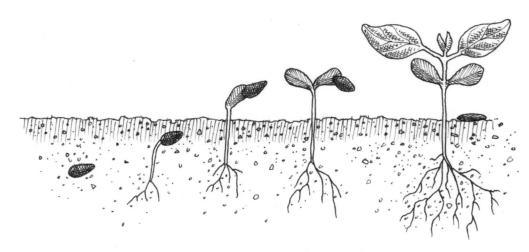

The most effective time to attack weeds is at the cotyledon stage. By the time they have more than two leaves, they are well anchored in the soil by their roots, and must be pulled out by hand.

right around the base of the plant stems without damaging the leaves.

In an ideal maintenance schedule, gardens are hoed every 10 to 15 days, especially during the months of June and July, when weeds are persistent and in direct competition with crops. It's also important to cultivate under good circumstances; if the soil is wet, any weeds hoed may re-root and the job will have accomplished little. The best time is on a dry, sunny day—and so that's when we plan this chore. In times of prolonged wet weather, it may be tempting to go out and cultivate anyway, but we've learned that it serves no purpose. One might as well advance other jobs ahead of time and be ready when the sun comes out. We've also learned to keep our hoes well sharpened. Doing so makes a big difference in terms of efficiency, especially when hoeing perennial weeds past the cotyledon stage. You want the hoe not only to disturb the weed, but to slice the roots of the uprooted plants. We sharpen our blades once a week with an electric grinder and carry a hand-held carbide sharpener (the best hoe-sharpening

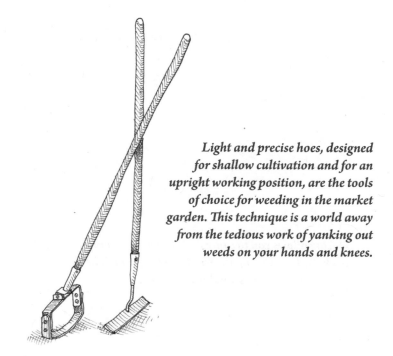

Light and precise hoes, designed for shallow cultivation and for an upright working position, are the tools of choice for weeding in the market garden. This technique is a world away from the tedious work of yanking out weeds on your hands and knees.

Hoeing allows you to control weeds while keeping a straight back. By breaking up the "crust" of the surface, hoeing also aerates the soil and stimulates plant growth.

tool we have come across) with us to the garden when cultivating.

Many people have trouble believing that weeding with hand tools is efficient and productive on a commercial scale, but we can vouch for this low-tech approach. With a good hoe and some practice, a market gardener can become very agile and swiftly cultivate the soil without damaging the crops. The flexibility afforded by manual hoeing is the main reason why we are able to space our crops, not according to standardized weeding implements, but however tightly we want to space them. The tools adjust to our practice, they don't define it.

But beyond the goal of eliminating weeds, spending time hoeing the crops also allows for an intimate contact with the soil and the vegetables. I find that this chore helps us get a good feel for what is happening in the garden, encouraging the development of botanical acuteness and sensitivity. Over many years, I have been able to watch vegetables go through every stage of development—and have learned a lot about plant biology in the process. All things considered, I believe hand hoeing not to be a step backward, but a simple and appropriate tool for the needs of a market garden. I have never envied vegetable growers who use mechanical weeding techniques, nor have I tried to find a better way to cultivate.

Weeding with Tarps

The main factor in keeping a garden weed-free is how much space is to be kept under control. Our 10 plots have a total area of 1½ acres, and if we had to cultivate the whole garden every week, I doubt we would manage. This is where the opaque UV-treated tarps comes in handy. Not only are they useful for smothering weedy ground and preparing the soil before planting, but when covering unused beds, the tarp limits the surface area on which weeds can establish themselves. Even more interestingly, we have also observed that black tarps do an especially good job of diminishing weed pressure on subsequent crops. The explanation is simple: weeds germinate in the warm, moist conditions created by the tarp but are then killed by the absence of light. This weeding technique, described as "occultation," is widely use by organic growers in Europe.

We have been using 6 mm black silage tarps in the garden for almost a decade now, and I can say without hesitation that their usefulness is one of the reasons behind the overall success of our operation. This passive and efficient practice takes care of part of the weeding chores while we are getting work done elsewhere in the garden. Besides being a petroleum product, the only difficulty we face using these bulky tarps is that they are heavy to move around. Our solution to this has been to buy more of them every year, with the intended goal of having one for every plot, thus eliminating the need to carry them from one garden to another. This minor challenge aside, the overall advantages far outweigh the drawbacks.

The Stale Seedbed Technique

The stale seedbed technique (also called false seed planting) consists of preparing seedbeds a few weeks prior to the seeding date, allowing weed seeds in the top 2" of the soil to germinate.

Soil-covering tarps are about as effective at fighting weeds as a short cover crop in a conventional vegetable rotation, but they can be set up in one go. They work immediately and quickly, and are the perfect fit for the intensive market gardening model.

When ready to seed, whether directly or by transplanting, the soil surface is then shallowly worked again, effectively destroying emerging weeds. The result is that the crop can get a head start over the weeds yet to come, and the difference between beds that have received the treatment and those that have not is striking.

For this technique to be effective, a few things must be considered. Firstly, it's important to let the beds go stale long enough for the weed seeds to germinate. In our gardens, we prepare them 10 to 15 days in advance and cover the bed with floating row covers. Secondly, the destruction of the emerging weeds must be done in such a way as to not bring up weed seeds that haven't germinated.

To enhance the stale seedbed technique, we use clear plastic row covers to induce weed growth. The results are very impressive—perhaps even scary, when you see just how many dormant seeds were lurking in the garden.

Soil Solarization

You can let the sun help you get rid of weeds by leaving clear plastic in place for 6 weeks during the summer. This is an effective strategy for depleting the seed bank and is a solution for weed management in hoophouses, or during early spring planting when stale seedbed techniques are not possible. Aside from requiring beds to be free during the growing season, the main drawback of this technique is that you don't destroy only weed seeds, but soil organisms and microbial life in the garden as well. Under certain circumstances, however, such an effective strategy is worthwhile.

A quick surface passage with our power harrow does a great job, but using a wheel hoe works just as well. To avoid any stirring of the soil and guarantee that buried weed seeds will not come up, a flame-weeder can also be used.

Since this simple technique gives tangible results, we try to use it as much as possible, especially for direct-seeded crops. To ensure proper results, we have systemized the practice by integrating bed preparation into our crop planning calendar. This isn't always possible for early spring seedings, and it doesn't always work out according to plan. For instance, double cropping doesn't always allow for an extra two weeks between both plantings, but adding it to our weekly schedules whenever possible certainly helps.

The stale seedbed technique is mandatory for our mesclun mix crop. When seeded at such a high density, this crop leaves almost no room for hoeing.* We therefore plan to seed this salad mix every two weeks on new beds, leaving us enough time to effectively stale the seedbeds. We pay just as much attention to our false seedbeds as we do to the mesclun mix since we want to ensure optimal growth in both cases. We irrigate the weeds as needed and protect them with floating row covers to keep the soil moist at all times. When you think about how much easier it then is to harvest a weed-free crop of mesclun mix, growing a thick carpet of weeds can be almost as satisfying as growing the greens themselves.

* We are presently trying out a hand-pushed finger weeder for cultivating dense stands of salad mix. Commercialized as the Grass Stitcher, this tool might just be the perfect companion for the 6-row seeder presented in the previous chapter.

Flame Weeding

Flame weeding is a technique in which weeds are killed by burning them with a blowtorch. Actually, the term "burning" is a bit misleading: the weeds are not burned to a crisp but rather subjected to a form of heat shock that causes damage at the cellular level. In order for flame weeding to be successful, two conditions must be present: the flames of the torch must actually make contact with the soil, and the weeds must be small enough that a single second of flame exposure is enough to kill them (i.e., between the cotyledon stage and the first true leaf). It's also important to flame over a very smooth seedbed, since irregularities in the surface of the soil can sometimes deflect heat from the flame weeder, giving some protection to weed seedlings.

Flame weeding nicely complements the stale seedbed technique. Burning the weeds avoids the need for the stirring action of a harrow, thus preventing buried seeds from being worked up to the surface. However, we rely on flame weeding mostly for burning weeds in the pre-emergence stage of direct-seeded crops. This approach is somewhat like the stale seedbed technique: seedbeds are prepared 2 weeks in advance to give weeds a head start, but instead of seeding after the destruction of the weeds, you seed into the stale bed halfway through the process. Then, just before the vegetable plants emerge from the soil, the flame weeder is run over the ground, leaving the crop to emerge into an essentially weed-free bed.

Pre-emergence burning is the ultimate way of providing weed-free beds for slow germinating crops that are direct seeded such as carrots, beets, and parsnips. However, this technique

Effective thermal weed control depends on the quality of the flame weeder. Ours is 30 inches wide and features five torches, which allow for burning at a high intensity with multiple flames. The torches are shielded from the wind by a metal hood—an important feature that allows us to use the tool in windy weather.

must be practiced with diligence. If you wait too long to burn the weeds and the vegetable begins to emerge, the vegetable crop will be totally invaded by weeds. If this happens, you will need to spend many hours weeding by hand or quite possibly have to reseed altogether, thus delaying the crop by a few weeks. To avoid missing the boat, we always throw on a handful of seeds that are known to germinate earlier than the main crop at the head of the seedbed. A patch of tiny radish plants signals that beets are about to pop up, while tiny beet tips indicate that the carrot crop is imminent. The appearance of the indicator crop tells us exactly when to flame weed so that we can do so with confidence. We also make sure to write a note on our crop calendar to remind us to check on the plants five days after the main crops have been direct-seeded. It's better to burn too early than too late.

Mulching

Mulching the garden with soil-covering material is another great way to keep weeds under control. Many home gardening books extol the virtues of using organic material such as straw, leaves, wood shavings, cardboard, etc. as ideal mulching material, but my experience leads me to advise against relying on them. Not only does plant-based mulch cover attract slugs, but weeds always seem to find their way up through them. This means you still have to weed—and weed by hand, because the mulch makes hoeing impossible. On a commercial scale, the cost and time required to spread thick amounts of organic mulch are too high, in my view. The one plant-based mulch cover that doesn't have all this inconvenience is grass clippings. We can easily produce it onsite (we mow our headways every other week), and its fine texture makes it easily digestible by soil organisms in the event that hoeing is required and we have to mix it into the soil. We've had interesting results working with ½-inch grass clipping mulches, but not to the extend of making it a systematic measure for weed control.

We've always found inorganic mulches to be more effective. We rely on both landscape fabric and biodegradable film to cover the beds of certain vegetables that stay in the garden for a long time, such as tomatoes, peppers, zucchinis, and melons. These mulches not only smother weeds, but also provide a highly beneficial environment for these crops, which prefer hot and humid conditions. Both products have their advantages and disadvantages.

Straw Mulch

Mulching with straw can be problematic. Most grain farmers who sell straw use herbicides to control weeds in their fields. This means straw mulch can carry herbicidal residues into your garden. Unfortunately, organic straw is hard to find in some areas—and if you find it, it may contain large amounts of weed seeds that you'll be importing into your garden. The straw we use to cover our garlic during the winter is made from first-cut fall rye. This straw is harvested at the beginning of the summer, when few weeds have gone to flower; it's therefore more likely to be clean and free of herbicides, even if produced by a conventional farmer.

Landscape fabric (we refer to it as "geotextile") is reusable and durable. Ours are in their sixth season and show little or no wear. They come in rolls of a width that will cover both beds and pathways—we use 16-foot-wide rolls that can cover 4 beds—and into which we've burned round holes according to the crop we intend to cover, in our case eggplants and melons. To make the holes, we use a small propane torch (it works best with a fine-tipped nozzle). We locate the holes by following a plywood template of the proper spacing. Burning more than 100 holes in such a way is a messy job (it almost made us reconsider this approach…), but once done, the work remains useful for many years to come. If landscape fabric is cut, the new ends should be burned to avoid unraveling of the fabric. Be sure to purchase a professional grade of landscape cloth, since the thickness of the fabric and tightness of the weave will affect how long it lasts.

The biodegradable plastic film that we also rely upon is a *lot* less expensive and more versatile, as it allows us to punch holes at any desired spacing. The one we use is 100% biodegradable and made of compostable genetically unmodified cornstarch resin. Since it leaves no toxic residue, we can mix it into the soil with a clear conscience at the end of the season. It comes in a 500-foot roll and is 36 inches wide, which is enough to bury the edge and still cover the width of our 30-inch beds. Although this plastic mulch is relatively fragile, we still prefer it to conventional plastic mulch, which ends up in the garbage after the crop has been harvested. Suppliers can be found in the tool section of this book.

Inorganic mulches can reduce the weeding workload to one or two early rounds of hand-pulling weeds. For crops that will stay awhile in the garden, the effort and investment required to lay them out it is definitely worthwhile.

The wheel hoe is an amazing invention. This tool makes it possible to hoe a large area without much effort. And unlike the two-wheel tractor we sometime use to harrow the pathways in our garden, the wheel hoe is sustainable and silent.

Weed Control Technology

I believe too many growers emphasize mechanical weed control as *the* solution to their weed problems. Most of the organic vegetable producers I know are always looking out for new cultivating tools: finger weeders, flex tine harrow weeders, torsion weeders, computer-controlled, and even robot cultivators (for real!). Upon first glance, it seems that having *all* of these tools or machines would be ideal, given that each one works precisely and effectively under very specific field conditions. Trade shows I attend are always lined up with growers eager to acquire new technologies for their farms. For the market gardener, such sophisticated tools are not currently available (at least not to my knowledge), and this might be a good thing as it enables us to look elsewhere for weed control solutions.

In this chapter I have laid out different techniques that help protect our gardens from weed invasions. Intensive spacing, transplanting, not turning the soil, never letting weeds go to seed, making sure not to import weed seeds from manure and mulches, and depleting the seed bank by stimulating germination are all solutions that are basically cost free. They do require, however, forethought and reflection in the planning and organizing stages of the cropping system. I firmly believe this is where a market gardener should concentrate his or her focus in order to effectively deal with the weeds in the garden. It's also important to understand and apply the difference between weeding and cultivating.

I was once at an organic farming conference where an experienced grower with more than 20 seasons under his belt was asked to name the top five most problematic weeds on his farm. After quickly naming two, he stalled for some time, causing an awkward silence in the room. After a few moments, he admitted that he didn't know the names of his weeds because he never let them grow big enough to make them out. I thought that said it all.

Insect Pests and Diseases

> *Insect pests do not exist: The moment the problem of crop disease or insect damage arises, talk turns immediately to methods of control. But we should begin by examining whether crop disease or insect damage exist in the first place.*
>
> — Masanobu Fukuoka, *The Natural Way of Farming: The Theory and Practice of Green Philosophy*, 1985

ANY SERIOUS DISCUSSION of plant protection in the context of market gardening must begin with the recognition that the use of synthetic chemicals designed to kill pests (animals, plants, fungi) is a catastrophe for the environment and for human health. The fact that these chemicals continue to be promoted, in spite of staggering evidence showing their adverse effects, proves that we cannot rely on industrial agriculture to nourish ourselves. The issue is immensely important, but I prefer to leave this discussion to others. For our purposes, we know these products are not acceptable in organic agriculture, while certain biopesticides are. But are biopesticides really any safer than synthetic pesticides?

It is also common to hear or read that parasitic infestations can be avoided in organic farming. The argument goes like this: when a grower pays attention to the biology, physical structure, and mineral balance of the soil, he can grow healthy plants that are naturally resistant to diseases and insect problems. If this is true, does it mean that the presence of such nuisances in a market garden is solely due to poor practices by the grower?

These are big questions to which I have no answers. What I do know is that certain insect pests and plant diseases appear in our garden every year and can cause major damage if we do not take preventative measures. Summer squash, for example, has little chance of surviving attacks by the cucumber beetle if left unprotected. I also know that customers are not willing to accept worm holes in their radishes or black rings around their tomatoes, just because the vegetables are organic. Therefore, managing pests and diseases is crucial to the success of our market gardening operation.

The first line of defense is biodiversity. The mere presence of diverse plants, insects, birds, and even amphibians living together on the same site

Covering vegetable crops with a protective net is an effective and environmentally sound strategy for preventing pest infestations. Anti-insect nets are highly resistant and durable. Unlike floating row covers, they have no effect on temperature, which is preferable for summer crops.

Ecological pest control means more than simply relying on biopesticides, which may or may not disrupt the spread of insect pests and diseases. It requires forethought and planning prior to the growing season. For every pest common to your growing area, you should know in advance an appropriate natural control and know when to intervene to disrupt the spread of the problem. Having prepared a guideline like the one on page 117 is a good starting place, but writing down the various management practices for dealing with them is just as important. For example, if you know that carrot rust flies usually emerge in August, write a note in your crop planning calendar to cover your carrots with a net around that time.

minimizes opportunities for pests to get out of control. The best way to encourage such biodiversity is to provide appropriate habitats for species that you wish to promote. Windbreaks or ponds may well include shrubs and other plants that attract insectivorous birds. Floral spaces created inside and along the outer edges of the gardens can be designed in such a way as to encourage insect predators; just as stone walls, trees, and other small refuges can encourage other beneficial insects. Many books discuss in detail such habitat management strategies, and these ideas are worth exploring when designing a market garden. A few such books are listed in the bibliography.

We have put a lot of effort into increasing biodiversity on our farm. Given that we began farming on bare fields surrounded by monocultures, doing so was a major undertaking. Our local beekeeper's beehives, located on the edge of the garden, have brought more pollinators to the site; and our pond is a haven for frogs, insects, and birds of all kinds. The presence of an aquatic environment, in close proximity to the woods at the edge of our farm, is beneficial to the toads that patrol our gardens at night to feast on cutworms. We have built several birdhouses to encourage bluebirds and wrens—important predators that hunt for insects in our soil. By steadily adding different ecological niches that benefit multiple species, we have managed to transform our market garden into a welcoming space for a wide variety of animal and plant species. Ladybugs, praying mantises, and lacewings are now a common sight in our gardens, signifying the farm's evolution. In most cases, all we had to do was integrate habitats into the garden landscape.

Having said this, the outcome of our effort is hard to truly assess, and we have needed other measures to avoid crop losses. Time and experience—including the experience of crop damage—have taught us to ward off plant diseases by doing targeted interventions at the right time. We've identified most of the pests in our area and have found specific solutions to control each one. In general, we favor physical controls (row covering, handpicking, pheromone traps, sticky traps, etc.) and use natural insecticides only as a last resort. Our specific management practices are described in detail in the crop notes in the appendices, but our overall control methods are pretty much the same throughout. One of the key elements is a speedy diagnosis.

Scouting

Almost all the pest management practices used in organic plant protection are preventive, which is to say they help avoid pest proliferation but cannot do much about it when a situation gets out of hand. For example, the natural fungicides that are permitted in organic agriculture will prevent a pathogenic fungus from spreading to new leaves, but will not actually get rid of the fungus itself. For their part, natural insecticides work best when used at specific stages in the pest life cycle. The leek moth, for example, can be effectively controlled with spinosad, but only when its larvae are creeping down the plant. When applying fungicides, insecticides, or both, timing is everything. Therefore, correctly diagnosing common pests and diseases is important. This is the purpose of scouting, which involves observing crops daily

We regularly inspect for tarnished plant bugs in our pepper beds. We do this by simply tapping the plants, causing the nymphs to fall onto a piece of white plastic, which makes them easier to identify. We intervene with a natural insecticide only if we detect the bugs in multiple locations in a single row.

and taking note of how potential risks might be developing.

On our farm, we have gotten into the habit of taking a walk around our gardens before starting work each morning. This little walkabout is brief, as the gardens are all close by. Yet even a short inspection allows us to detect any anomalies that might be present in our plants and determine where to allocate our maintenance efforts. It's helpful, however, to know what you are looking for, which is why we subscribe to a phytosanitary alert service that tells us which diseases and insects to watch out for. This helpful email service is

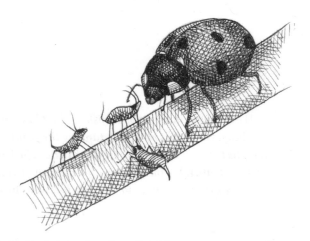

We regularly introduce beneficial insects to our greenhouses to drive out insect pests. So far we have had success in controlling thrips by introducing predatory mites. We also purchase ladybugs if aphids become a problem.

free of charge. It features regional alerts based on the expertise of professional researchers and crop scouts. Considering the amount of vegetables we grow, this is a service I would not want to be without. We also have a number of reference books on the main crop pests, their life cycles, and different techniques for spotting them. Suggested books are listed in the bibliography.

Disease Prevention

A stretch of overcast, rainy weather for several weeks is a sign that garden diseases are likely to proliferate. The Solanaceae and Cucurbitaceae families are especially susceptible to infectious pathogens, but other vegetables such as onions, beans, and greens can also suffer if the weather is unfavorable for too long. Depending on the strain of the infection, plant diseases can lead to dysfunctions and eventually crop losses. Whenever we detect a vegetable disease, we first try to determine the pathogen in question. In most cases, the crop leaves will exhibit some kind of symptom (blotches, wilting, burning, yellowing, necrosis, etc.), which we can compare with a guide to identify the culprit. This is not always easy to do, as multiple diseases may be present on a given leaf at the same time, and certain symptoms are easy to confuse with physiological problems (deficiencies, moisture stress, etc.). This is where phytosanitary alerts are especially useful: unlike books, the diagnosis is based on current weather conditions and is therefore probably more accurate. There are three main vectors of vegetable disease: viruses, bacteria, and fungi. These are key points to know about each of them.

Viral diseases are the least common, and so far our farm has been virus-free (touch wood!). Most viral diseases spread via seeds, so it's important to use high-quality seeds. Garlic seed is particularly at risk, as are tomato transplants from commercial greenhouses. This is one good reason to produce your own seedlings.

Bacterial diseases can be surprisingly severe and require rapid intervention. They usually manifest in rainy weather. Affected plants will generally find themselves close to each other, as bacteria propagate through physical contact. Getting rid of the disease involves removing sick plants from the soil, taking special care that they do not touch any other plants in the process. On our farm, we throw diseased plants in the garbage and even change our clothes after working with them. One of the symptoms of bacterial disease is rot, which quickly causes the crop to wilt. The bacterial wilting that affects our Cucurbitaceae crops is the only bacterial disease that returns every year to our farm.

Fungal diseases are relatively common. The first precaution to take is to avoid damaging plant tissue at all times (from weeding to staking to harvesting), as fungi require a point of entry into the plant. When pruning, it's also really important to work on a sunny day, as water in an open wound creates a perfect setup for the appearance of diseases such as blight.

If we detect a fungal disease, the best we can do is to contain the damage by making weekly sprays of copper and sulfur in alternation. These sprays are not curative but do allow the plants to continue growing if applied in a timely fashion. These treatments are effective, but there are some

Not all backpack sprayers are of equal quality, and it's worth spending the extra money to get the best one available. We have been using the same one for many seasons, and because we take the time to rinse it after each use, it still works very well.

drawbacks: copper can accumulate in the soil and inhibit biological activity, while sulfur is harmful to garden insects. For these reasons, we are now exploring different biological fungicides that introduce bacteria to fight certain pathogenic fungi. We are also looking into inoculating the beds of susceptible crops with beneficial microbes that will protect plants from soilborne pathogens. Both these practices, unlike mineral-based fungicides, supplement rather than harm biological

Completely sealing our hoophouse with insect nets protects our cucumbers against cucumber beetles and the bacterial wilt they propagate.

Using Biopesticides

Insecticides are the last resort when it comes to insect problems on our farm. We have a love-hate relationship with these products: we know they are not harmless, but at the same time, they are essential for the protection of certain crops under specific circumstances. I strongly believe that the market gardener should strive for ideal practices, but not at the expense of losing a crop. Using biopesticides to address specific needs is a good example of this.

The prefix "bio" in front of "pesticide" indicates that these products are natural in origin (not chemically derived), break down over time, and do not pollute the soil with toxic residue. Bio-pesticides may be persistent (active for several days), selective (intended to attack a specific host), or broad-spectrum (active on a number of insects). In all cases, they are toxic and should never be viewed as harmless substances just because they are allowed in organic agriculture. Products such as pyrethrum and spinosad are very potent agents that must be handled with care, especially when mixes call for concentrates. It's also important to keep in mind that most biopesticides are manufactured by the same multinational corporations that manufacture and promote the synthetic pesticides that damage human health and the environment. Claims that the safety of these products has been tested can therefore be met with justifiable suspicion. Rotenone, until recently a commonly used biopesticide, is a good example of this. It was permitted by organic standards, but later banned due to its association with Parkinson's disease in farm workers who handled it regularly. This example is a reminder that when

life in soil. Experience has taught us to pay close attention to what the seed catalogues have to say about the different cultivars: some are resistant to diseases that we have encountered in our gardens. These seeds are more expensive, but they often stop the problem at the source. Relying on resistance in the varieties is especially important when it comes to growing in greenhouses and hoophouses. In these moist milieus, tomatoes and cucumbers are not rotated and are thus highly susceptible to annual contamination.

dealing with such potent agents, wearing protective clothing (gloves, goggles, etc.) is important. It's also good practice to make a chart detailing the proper dosage calculations for each product, along with a reminder of the recommended amounts of time between sprayings.

As a parting comment about biopesticides, I wish to emphasize that we use insecticides on our farm not to kill pests but rather to reduce crop damage by controlling their population. Below is a table of our preferred interventions to fight certain garden pests. These references should be used only as a guide, as they will need to be updated periodically in the future. They are specific to our site, and it should be noted that there are other pests which we have not yet encountered or which have not caused us enough damage to justify an intervention.

Solutions to Fight Insect Pests at Les Jardins de la Grelinette

	Anti-insect netting*	Btk**	Handpicking	Insecticidal soap	Kaolin clay	Orthophosphate	Pyrethrum	Spinosad
Aphid				P			O	
Cabbage maggot	P							
Carrot rust fly	P							
Colorado potato beetle	P		O					O
Cutworm		O	P					
Flea beetle	P						O	O
Leek moth		O	P					O
Mustard white		P						O
Slug			O			P		
Striped cucumber beetle	P		O				O	
Swede midge	P							
Tarnished plant bug	P						O	O
Thrip				O			P	

* Mesh size must be chosen based on insect size.
** *Bacillus thuringiensis* var. *kurstaki*

P = Preferred solution   **O = Also effective**

Season Extension

The intelligence of market gardeners has been applied to means of forcing nature to produce in the middle of winter, in the middle of the frost, what is ordinarily produced only in the warm spring and summer weather; it is in this area that the science of the gardeners of Paris has revealed itself to be truly astonishing.

— *Manuel pratique de la culture maraîchère de Paris*, 1845

OUR GROWING SEASON in Quebec is short, and we therefore must take every means necessary to produce as much as possible in the little time we have. In this climate, our task as market gardeners is to find ways to control the growing conditions, in order to protect our crops from the cold and frost in both early spring and late fall. In this regard, I should clarify at the outset that when I talk of "season extension" in this chapter, I am referring to using unheated and minimally heated tunnel-like structures along with a whole set of simple and economical technologies and techniques that help growers force crops and defend them against harsh weather.

The idea of forcing crops to grow out of season is certainly nothing new, but doing so usually equates to heating greenhouses. In the last 50 years—during the era of cheap fuel—high-tech approaches to greenhouse production evolved tremendously, but so did the complexity and cost of these systems. In the northeast United States, and other regions where demand for local products is increasing, many small growers have set out to find simpler, less energy-intensive ways to get earlier harvests and extend their season into the winter. These exciting innovations are becoming better documented, and the body of knowledge about using hoophouses for cool-weather crops has grown each year. At the forefront of this trend is Eliot Coleman. His ideas and techniques have helped growers push the extent of their season by combining an understanding of plant biology and low-tech cover solutions. I have seen these methods at work with my own eyes on a number of cold climate farms and can assure skeptics that diversified production from October to May is possible. For those interested in year-round, cold climate production, there is a whole body of literature out there on the subject.*

* Eliot Coleman's *The Winter Harvest Handbook* is a comprehensive manual on year-round vegetable production. Other American growers such as Steve Moore and Paul Arnold have also written extensively about their successful experiences.

In the 19th century, French market gardeners used cloches and cold frames to force vegetable growth during the cold season. Today we rely on floating row cover and polyethylene film to do the same thing.

methods we rely upon are all passive. These techniques allow us to produce not just earlier vegetables but better vegetables—and more of them. The improved quality and yield are in my view the best reason to adopt season extension techniques.

Floating Row Cover and Low Tunnels

In my opinion, floating row cover is one of the all-time greatest technological innovations in the horticultural industry. It's made of webs of unwoven polymer fiber, a material that lets air and water through while still acting as a physical barrier against wind and insects. When spread over crops, row cover increases the soil temperature by 2–3 degrees while also keeping in moisture, thereby protecting the plants against frost. Installing row cover over direct-seeded crops speeds up germination; installing it over transplanted crops protects the seedlings from harsh weather such as driving rain, strong wind, and hail. Essentially, row covers can create a microclimate anywhere in the garden, comparable to that of a hoophouse.

Row covers are available with thicknesses expressed in ounces per square yard. The thicker sheets have a higher thermal capacity but block more light. On our farm, we use 0.55 oz/ft^2 (19 gr/m^2) row covers in spring and fall. These are a suitable compromise between durability and light transmission (they obstruct about 15% of the rays). We also use heavier ones, which are about twice the thickness, over fall crops to act as a thermal screen on frosty nights.

In the spring, our entire production is under row cover. For direct-seeded crops, we spread the

On our farm, for time being, we have decided not to grow vegetables throughout the year. It is not the technical challenges that are holding us back; we just really look forward to taking some time off in the winter. Having said this, we still do force crops, especially in the spring. Over time we've come to the conclusion that people are more excited about vegetables in early summer, and our goal is to maximize harvest in June. To do so, we don't hesitate to use propane gas to heat our tomato greenhouse, but other than that, the row cover directly on the ground, leaving a little slack to give the crops some room to grow. For transplanted crops, which are more fragile, we support the floating cover with hoops made of 9 gauge galvanized steel merchant wire. The wire is cut into five-foot lengths to create semi-circles that straddle our 30-inch beds with enough inside clearance for most crops. The hoops are pushed into the soil and spaced every 3 feet along the row.

For broccoli, summer squash, and other crops that grow tall, we use larger hoops made of ½" PVC plumbing pipe (PEX type) cut into eight-foot lengths. We space them every five feet or, if the same cover is shared by multiple adjacent beds, every 10 feet in a zigzag pattern. The hoops are driven into six-inch holes we make with a stake.

Finally, we use other hoops that we refer to as low tunnel hoops. These are especially useful when snow accumulation is still a possibility. They are made from 10-foot-long galvanized electrical conduit (EMT), bought at a local hardware store and bent into shape using a hoop bender. These low tunnel hoops are more expensive to make, but they are sturdy enough to handle heavy snow loads. In early spring or late fall, when snow accumulation is still a possibility, we cover our low tunnels with clear greenhouse film. The low tunnels then offer basically the same advantages as hoophouses but at a fraction of the price.

To hold down our row covers (both polymer fiber and plastic film), we bury the edges with dirt taken from the pathway, or if available, we use sandbags placed at the bottom of the hoops. To ensure that the bags will last a long time, it's worth investing in the UV-treated bags. When in-

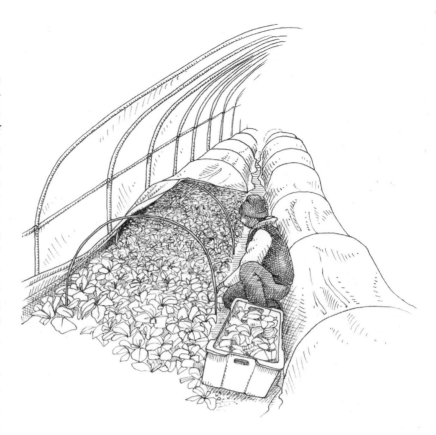

For three winters, we grew spinach, a very cold-hardy crop, in an unheated tunnel protected by floating row cover at night. Baby spinach could easily be called the queen of winter crops.

stalling row cover, we always make sure to stretch it tightly so that the fabric does not beat in high winds. Unfortunately, row cover is always at risk of being whipped up in the spring on windy sites like ours. The only solution we have found is to check regularly to make sure the weights are doing their job.

When we are done with the row covers, we stuff them into old grain sacks which we label according to the width and condition of the row

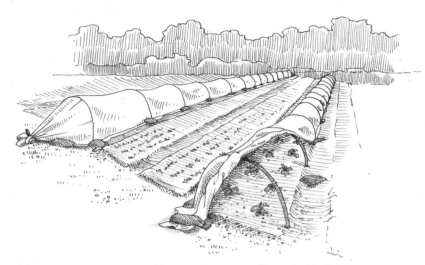

Floating row covers are effective, easy to install, and affordable. We attribute the success of many of our crops directly to these materials.

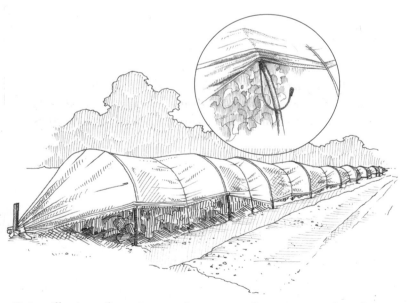

Caterpillar tunnels are inexpensive compared to permanent hoophouses. They provide shelter, warmth, and air flow to the growing environment.

cover. At the end of the season, one of the last chores we do is to try and patch up the ones with holes using UV-treated sheeting tape. Most of our row covers last approximately 3 years.

Caterpillar Tunnels

Caterpillar tunnels are another great option for season extension. In short, a caterpillar tunnel is an inexpensive and simpler variation of a hoop-house, with the interesting feature that it is mobile. Because its structure is easy to set up and take down, it can be moved over established crops at different times in the season. We use ours to force early direct-seeded carrots and beets, and then move it over our Solanaceae crops, which always love the extra heat in the summer.

Caterpillar tunnels can be built using any number of techniques and materials. The more basic option requires nothing more than PVC pipes, rebar, and rope. We built ours with 1½" PVC piping, made up into 20-foot lengths using two pieces joined together firmly with a glued union. The hoops are spaced every 10 feet and anchored to the ground on 24-inch rebar (⅝" diameter) driven halfway into the soil. We tie a rope from hoop to hoop to form a purlin that runs the length of the tunnel. The rope is then firmly stretched to give rigidity to the structure and tied to stakes at both ends of the tunnel. The covering (we use old greenhouse plastic) is then pulled over the tunnel and held to the ground by sandbags. To keep the plastic from flying away on windy days, we stretch rope over it midway between each arch and tie the rope down to stakes in the ground, giving the tunnel its caterpillar-like appearance. Ventilation is controlled by simply rolling up the entire side of the tunnel, keeping it in place by hooks attached to every other arch.

In 2005, we spent about $400 in materials, not including the used greenhouse film, to make a caterpillar tunnel that covers an area of 10 feet by 100 feet. Providing cover for 6 beds (the tunnel covers 2 beds and is moved 3 times in the season) for an investment of $.33/ft² is more than a bargain. The only downside of caterpillar tunnels is that they are low structures, providing little clearance to properly work in an upright position. Entering and leaving the caterpillar also involves a lot of bending.

Hoophouses

A hoophouse (also referred to as a polytunnel, high tunnel, or tunnel) is a permanent structure made of semicircular steel hoops bolted down and covered with a plastic film. Unlike greenhouses, hoophouses are relatively low, easy to build, and usually not heated. The hoophouses I mention here are different from the big tunnels commonly used in large-scale vegetable operations: they are generally 16 or 20 feet wide and approximately 10 feet tall at their peak. A hoophouse can be built to any length and even extended by adding new sections—a useful feature for beginning growers who may expand operations in future years.

One of the main advantages of hoophouses is that they can be used all year long.* These shelters can be used for early spring and late fall crops, before and after heat-loving summer crops. If seeded

* In very snowy areas, each hoop must be supported with a post.

Since caterpillar tunnels are portable, we can reap all the benefits of keeping crops covered without compromising our crop rotation.

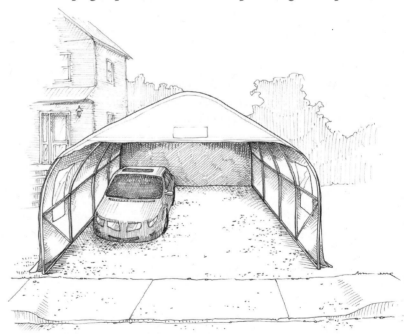

Car shelters are easy to find anywhere in Quebec. You can easily attach them together to make an affordable hoophouse.

We currently use hoophouses for growing trellised cucumbers and peppers in the summer. Prior to erecting our greenhouse, we grew our tomatoes in them as well.

Single or Double Poly?

A double layer of polyethylene inflated with air using a blower greatly increases the insulation factor of greenhouse-type shelters. However, having two layers of plastic also reduces the amount of light that can reach the inside of the shelter. When growing in hoophouses, it's probably better to install only one layer of plastic, maximize heat gain, and insulate crops as needed using row cover.

at the right time, cold-hardy vegetables such as spinach and Asian greens may be harvested even in periods of frost. Hoophouses come in many models and are available from most greenhouse suppliers. Since new structures can be expensive—and the price of metal seems to go up every year—many small producers decide to make their own by bending steel pipes with a metal bender. In my opinion, however, the best way to reduce the cost of a hoophouse is to look for a used one. Although this often means dismantling someone else's old structure and transporting it yourself, it's often worth the trouble. Whichever way you decide to go about it, building a hoophouse is one of the best investments a market gardener can make. Sales of the crops grown in such a protected environment will pay for the structure within a few seasons, if not the first.

Because hoophouses are permanent structures, they must be positioned carefully. You must ensure proper soil drainage beforehand so the beds are dry enough in the spring. Apart from leveling the ground, which requires heavy work in the garden, the best drainage solution is to install weeping tile around the structure. Our hoophouses have roll-ups and big doors at either end, providing the growing area with sufficient ventilation.

Season Extension Preferences of Certain Crops at Les Jardins De La Grelinette

S = Spring SU = Summer F= Fall

Vegetable	Row covers or Low tunnel	Caterpillar tunnel	Hoophouse	Greenhouse
Arugula	S			
Asian greens	S&F			
Basil			SU	
Beet	S&F			
Broccoli	S			
Spinach	S		F	
Carrot	F	S		
Cauliflower	S			
Cucumber			SU (trellis)	
Eggplant		SU		
Kale & Swiss chard	S			
Kohlrabi	S			
Lettuce	S&F			
Melon	S			
Mesclun mix			S&F	
Pepper			SU	
Radish	S			
Summer squash	S			
Tomato				S
Tomato Heirloom		SU		
Turnip	S			

Harvest and Storage

Knowing how to grow is not everything. One also needs to know how to harvest so as not to compromise, so close to one's goal, that which is the fruit of so much effort, expense, and care... The diligence with which this work is done is therefore of great importance.

— Abbot François-Xavier Jean, *Les champs. Manuel d'agriculture conçu par les professeurs de l'École supérieure d'agriculture de Sainte-Anne-de-la-Pocatière*, 1947

HARVEST IS TRULY one of the highlights of the season. This is when all of our efforts and attention to detail get translated into a tangible finished product. As a grower, nothing makes me prouder than harvesting a successful crop. Having said this, harvest requires some special know-how. To ensure quality is maintained all the way from field to fork, certain principles must be understood and followed through.

First of all, it is important to harvest each crop at the right stage of growth. Vegetables harvested before they are fully mature will be much less flavorful; those harvested overripe will have a shorter shelf life. For some crops, it is easy to tell when they are at their freshest, and many can be quite forgiving. But for other crops, harvesting at the opportune moment will make an enormous difference. The timing of cantaloupe picking, for example, will determine whether the fruit is sweet or disappointing. Each vegetable has its own set of signs to watch out for. I have listed the most

notable ones we watch for in the crop notes of the appendices. To further complicate things, the right moment to harvest a given vegetable does not necessarily coincide with market demand, and this is one reason why a cold room is indispensable in a market garden. Vegetables such as broccoli, summer squash, and cucumbers, to name just a few, can be harvested at the peak of their quality and stored in the cold room for a few days before being sold.

The other important thing to remember is that vegetables go on "living" even after being picked. They must be cooled to slow their respiration, otherwise their freshness and nutritional value will suffer. This is why it is so important to harvest as early as possible in the morning, before the heat of the day sets in, and to cool the vegetables immediately with cold water and/or air from the cold room.

Post-harvest procedures on our farm are pretty straightforward, since vegetables are not

When incoming crops from the garden are waiting to be washed, it is important that they be pre-refrigerated. Our well-insulated storage area is always about 60°F, which is quite suitable. You can also cool harvested vegetables by covering them with a soaked wool blanket.

stored for very long. We sell almost all of our vegetables the day after they are harvested. We keep our storage space at about 60°F (15°C), which is cool enough for most vegetables. Those that need to be refrigerated are kept at 36 to 39°F (2°C to 4°C) in the cold room. The basic harvest process is pretty much always the same and can be divided into three steps: collection from the garden, temporary storage and washing, and refrigeration in the cold room. However, there are some exceptions that require special attention.

LEAF VEGETABLES, including lettuce, are always the first to be harvested and are brought to storage immediately in harvest bins. As soon as they arrive, the leaves are sprayed with cold water to keep them cool until the washing step, which is done later in the day. The washing step involves carefully removing damaged leaves, submerging the vegetables in a cold water bath for a few seconds, and leaving them to drain until they have all been dipped. The vegetables are then gently placed in a closed bin and stored in the cold room.

ROOT VEGETABLES SOLD IN BUNCHES are sorted and bunched in the garden or, in hot weather, brought to the cool storage area in bulk to be bunched. Any damaged leaves are removed during the assembling process, and the bunches are calibrated to ensure uniformity. (The desired number of bunches is determined beforehand and the corresponding number of elastics counted out—e.g., 40 elastics for 40 bunches.) While awaiting their turn at the wash station, the root vegetables are kept in harvest bins and sprayed with cold water so that the tops stay fresh. At the washing stage, any dirt is rinsed off the roots with a pressure hose. The clean bunches are then placed in bins in a staggered fashion. Care is taken not to overload the bins, which are taken to the cold room.

BROCCOLI AND CAULIFLOWER will stay crisp longer if they are cooled immediately after being harvested. As soon as they arrive at the storage area, they are immersed in a cold water bath, drained and taken immediately to the cold room. Since broccoli and cauliflower heads are not picked systematically on harvest day but rather as they reach maturity, it is important to clearly mark the date of harvest on each bin so that they can be distributed in order of freshness.

BEANS AND PEAS do not need to be washed, but if harvested midday, they benefit from being sprayed with cool water before going into the cold room. In this case, you must ensure that they are put in a container that allows the water to drain out properly, in order to prevent rust—especially for beans.

CUCUMBERS will be crunchiest if they are cooled after being harvested from the tunnels. The freshly picked cucumbers are immediately immersed in a cold water bath, drained, and stored in the cold room in a bin marked with the harvest date.

TOMATOES may be harvested at any time of day, but they must always be handled with great care, as damaged tomatoes have a shorter shelf life. To avoid handling them twice, we store tomatoes in the same bins used to harvest them and keep them at the ambient temperature of the storage area.

MESCLUN MIX is often harvested on a different day from the rest of the crops so the entire cutting can be done early in the morning—it can be a very long job! Once harvested, the greens are immersed in a cold water bath and gently swirled to mix up the different sizes and colors. Any weeds, insects, or damaged leaves are removed at this stage. The mix is then spun in an electric spinner to prevent rot and delicately packed into sealed plastic bags, which go into the cold room.

MELONS, like tomatoes, can also be harvested at any time of day. They are usually not refrigerated but instead left to ripen at the ambient temperature of the storage area. If they are harvested too ripe by mistake, however, they are taken to the cold room, even though this reduces their flavor somewhat.

BASIL can be harvested at any time of day, but it must never be harvested wet or kept in a sealed bag, as moisture will turn the leaves black. Basil is stored in the cold room in a half-open bin to prevent condensation.

SUMMER SQUASH is harvested every two or three days when the fruits are still small. The fruits are not washed. They are put in the cold room in bins marked with the harvest date.

SUMMER ONIONS may be harvested any time of day and are often saved for when most of the harvest is finished. They are bunched in the garden, hosed down to remove any soil, and stored in the cold room.

Harvesting Efficiently

In the mid-1940s, Abbot François-Xavier of La Pocatière used the word "diligence" to explain that harvesting must be done quickly and efficiently. He was right. Harvesting is a task that can drag on and on, and if you spend too much time on some vegetables, others will wilt before they are picked. Drooping vegetables can always be perked up again in very cold water, but part of their quality will be lost. When it comes to harvest, time is of the essence—more so than with any other farming task.

Aside from having extra help, the best way to keep the harvest moving along is for the grower to master harvesting techniques that enable maximum efficiency. Since most harvesting involves small repeated actions, it is important to take the time to study the ergonomic aspects of these movements. By dissecting the minute motions involved in harvest, one can discover how to do the job while minimizing extraneous effort. This

When harvesting we try to pick as much as possible in one trip. To make sure that the vegetables stay shaded, we installed a removable umbrella on our harvest cart; and we often carry a damp wool cover with us to help keep the vegetables cool.

Harvesting implements can really speed up the harvest. Farm visits are a great way to make such discoveries, but don't neglect to also use your own imagination. Many pieces of equipment on our farm have been custom-built to fit our exact needs.

requires practice and a certain amount of self-awareness, but taking the time to perfect picking techniques can save hundreds of hours of unnecessary work in the long run.

Another important aspect of efficient harvesting is to organize the workflow in order to minimize the number of times the vegetables are handled. Finding ways to cut down on trips between the garden and the storage area is especially important. For example, in order to avoid having to fetch more elastic bands in case we miss some, we always carry an extra box in our harvest cart. We do the same with field knives, which are stored in a tool box fixed to the cart. The main thing to remember is that taking care of tiny details now can save huge amounts of time later. I often tell our interns that efficiency is a state of mind that can be reached only by conscious effort.

Harvest Help

In our world of market gardening, harvest is the most labor- and time-consuming part of the job. If there is one area where outside help can be useful, this is it. Laborers are often people visiting the farm temporarily, such as enthusiastic CSA members, customers, or travelers who work for a few weeks in exchange for room and board. They can also be part-time employees for the season. For more than a decade, we have welcomed woofers and visitors, and although we like having youthful travelers with us, I will say without hesitation that a paid harvest laborer, with training and experience, is much more profitable, even if volunteers are not paid. Be that as it may, certain measures are necessary in either case.

First and foremost, with inexperienced workers of any kind, never make the mistake of assuming that what is obvious to you is obvious to them. (If you do—be prepared for time wasted or harvest lost.) In my time, I have seen volunteers cut leeks off at the base, and people "harvesting" peas by tearing whole plants out! The stories we could tell.... What we have learned from this is to never leave inexperienced harvest helpers unsupervised. Being by their side all the time is educational for them (this is why we host interns on our farm), but having to explain every step and watch their every move often makes our work more time-consuming and less efficient than it would otherwise be.

With permanent employees or interns who will stay for at least a few weeks, it's a different story. We then take the time to train these people so that they can work unsupervised, thereafter allowing us to harvest multiple crops at once. But even then, regular check-ins are required to make sure they are doing the job right. Bunch size, which tends to fluctuate over time, is a classic case in point. To try and standardize our working practices, we post clear instructions in bold print in our work spaces to help guide our helpers. This is a simple measure that can make harvesting much more efficient.

The Cold Room

As I have already mentioned, a cold room is an invaluable asset for any market gardener. The cold room has three purposes: to cool incoming vegetables from the garden by forced air, to store them for extended periods, and to chill them to a

Our cold room allows us to harvest the day before delivery. This takes a lot of stress off our backs and saves us from getting up at the crack of dawn on harvest days. However, storing vegetables in cold rooms does require a certain logistical aptitude.

point where they will stay fresh during transport, regardless of the ambient temperature of the delivery vehicle. The type and size of the refrigerated space are very important factors in accomplishing these aims.

To anyone setting up a cold room, the two best pieces of advice I can give are to invest in a new air compressor with a warranty—refrigerator repair people charge about $100 an hour—and to set up a larger cold room than you think is necessary. It is not overkill to get double the capacity needed for your current level of production. It may seem counterproductive to refrigerate all that space you won't fill, but it is better not to limit yourself by underestimating your future needs. Also, a generously sized cold room comes with certain advantages that should not be overlooked.

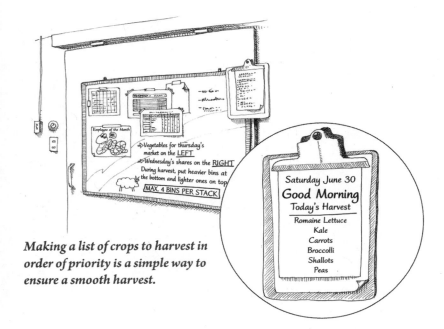

Making a list of crops to harvest in order of priority is a simple way to ensure a smooth harvest.

Firstly, a large cold room with a powerful compressor will be better able to handle the incoming heat from frequent trips in and out at harvest time. The best way to bring the vegetables to their ideal storage temperature is to have a cold room that actually stays cold.

Secondly, having a refrigerated space that is not filled to capacity will allow air to flow more freely—and air circulation plays an important role in cooling vegetables. For this reason, it is a good idea to allow for a four-inch space between stacks of bins.

Lastly, it is much easier to manage the contents of your cold room when you have a large space to work in. Different sections of the cold room can be devoted to specific vegetables, and it is much easier to maneuver a cart full of bins when you have extra space. Our cold room at Les Jardins

de la Grelinette measures 8 feet by 16 feet—and it was only in our fifth growing season that we began to occasionally fill it completely.

When it comes to organizing your refrigerated space, choosing the right storage bins for your vegetables is important. This involves shopping around, as there are many different models available. Ideally, storage bins should have the following features:

- They should be the right size: not too small, but not so big that they are too heavy to carry when full of vegetables. It may be a good idea to have three different sizes of bins: one for leaf vegetables, one for root vegetables, and one for fruit vegetables.
- They should have a closable lid to hold in moisture and prevent vegetables from drying out in the cold room.
- They should be nestable and stackable. This means they need to be able to support heavy loads when stacked and be sufficiently robust to last many seasons of intensive use.
- They should be easy to clean and have holes in the bottom to allow wash water to drain out.

So far, we have experimented with various kinds of bins, new and used, and even customized them, but I have found that they all have different features that should be rolled into one product. Alas, the quest for the perfect bin continues. Sometimes we run out of bins, either because we have an especially plentiful harvest on our hands or because we have a lot of vegetables in storage. In this case, we use harvest bins; these do not have lids but significantly limit moisture loss when stacked.

Crop Planning

*Give me six hours to chop down a tree and I will
spend the first four sharpening the axe.*

— Quotation attributed to Abraham Lincoln

ORGANIZATION = SUCCESS. For the market garden, this adage is especially true at the production planning stage of the job. This process can be particularly difficult, however, when you have little or no experience as a mixed vegetable grower. It was for this reason that I decided to cover this topic at the end of the handbook, even though crop planning really comes at the beginning of any growing season.

Crop planning is absolutely fundamental to profitable market gardening, and we owe a large part of our success to our skill in this area. Knowing *what* to grow, *how much* to grow, and *when* to plant is not a simple task, especially when precision is required. Doing so requires, above all, an understanding of the steps of crop planning. It might seem confusing at first, but hang in there, it follows a straightforward logic. Then it becomes a matter of following the process through with rigor. The planning stage can be an overwhelming endeavor, but it is worth every effort put into it. All that is carefully planned and anticipated for the year ahead enables the streamlining of many operations during the season. Winter, when more time is available, is the perfect time to do this.

Setting Farming Objectives

We begin our crop planning sessions after taking a few weeks of vacation away from the gardens. Just as taking a power nap makes for great working afternoons, taking a leave from our daily routine helps us freshen our mind for the task at hand. Our first step is to come up with the annual budget. Setting financial goals is a priority for us because we farm to support our family, first and foremost. Therefore, we must make sure our production will generate targeted revenue, or at least one that we can make do with. Our financial objectives are then translated into sales objectives, which in turn determine our production objectives for the season. While this sequence of priorities may seem obvious to some, many farmers do it backwards, establishing their production capacity first and afterwards hoping to make ends meet. I strongly discourage beginning farmers from working that way. My father would always say, "A goal without a plan is just a wish." If you want to make an adequate income by market gardening, it is best to plan for it.

In the CSA model,* production objectives are expressed in the number of shares, their weekly price, and the number of deliveries. For example, producing 60 weekly shares at $23 for 18 consecutive weeks generates about $25,000 in revenue. Determining these numbers will depend on the available growing space, the work force, and especially the grower's experience—all of which need to be considered. This is the step that determines the scale of a market gardening operation.

The second step is to decide the type and quantity of vegetables to grow in order to reach these production objectives. This step is the more challenging one and is done in two parts: first, you need to roughly decide the contents of the CSA shares for every week of the season. You then need to calculate the planting dates in order to harvest these crops on time and in the quantity that will be required to meet the demand. The tables given in this chapter are very useful for this planning.

The third step is to regroup every seeding by species, making sure that everything fits into the garden space available. With all of this information in hand, you then develop your crop calendar and a garden plan. Once these two are established, running a complex production system becomes much simpler.

Finally, the last step of crop planning is an on-

* Production for sale at farmers' markets can be calculated in just the same way as when planning for CSA production.

going one, which is to keep records throughout the growing season, taking notes of what went wrong and/or of what could have been planned more precisely. These notes will be of crucial importance the next time around, when you prepare your crop plan for the following season.

The method described here is by no means universal. Many farmers do it differently, but this process is coherent and is relatively simple to follow. Using our own farm as an example, let's now go through the whole process one more time, but in detail.

Determining Production

At Les Jardins de la Grelinette, we now produce 120 CSA shares over 21 weeks and sell vegetables at two farmers' markets for 20 weeks. Since it is hard to determine in advance what we will be able to sell at the farmers' market (as this is subject to weather, how busy the market is, etc.), we estimate that our market customers equate to 100 extra CSA families. Admittedly, this estimate is a bit of a cheat because the market customers do not necessarily buy the same vegetables as those distributed in the CSA shares, but the demand is similar enough that we can base our planning on this assumption. This also happens to be the simplest way we know of going about it.

So, in total, we plan to produce 220 shares every week, each with an estimated value of around $26 worth of vegetables. This targeted revenue of $117,260 (220 shares × $26 × 20.5 weeks) is enough to meet our financial objectives, allowing us to enjoy our chosen lifestyle.

What to Grow

Next, we need to get more specific about what to grow. To do this, we create a table with 21 rows, corresponding to our weekly deliveries. We then define the shares by making a list of assorted vegetables along with their values. These vegetables are chosen based on seasonal availability, but also on subjective preferences such as crowd-pleasers, or crops we like to grow (or don't like to grow). In the process, we pin down the exact contents of the first three and last four shares of the season. These ones are the most critical given the limited availability of crops we are able to grow at those times.

For shares 4 to 17, we don't calculate weekly contents as precisely, but rather make a rough plan of all the vegetables that will be grown in succession. When the season is in full swing and many vegetables are ripening all at once, shares are built with some vegetables that will store (either on the plant, in the ground, or in the cold room) and some that must be harvested and sold quickly (e.g., peas, beans, tomatoes). We calculate our needs differently according to the type of vegetable.

For single-harvest vegetables (roots, lettuce, broccoli, celery root, etc.), we determine the number of times we want to include these in our shares (e.g., 8 times for carrots, 5 times for beets), which gives the number of seedings we will have to plan for. For multiple-harvest vegetables (tomatoes, cucumbers, summer squash, etc.), the objective is to plant enough to produce 220 units per week. For summer squash, for example, this means having 110 summer squash plants producing at a given time, considering that a single plant gives an average of two fruits per week.

Here is what the planned shares might look like for a given year:

SHARE 1 (June 13): spinach ($3), radishes ($2), cucumbers ($4), summer squash ($4), kohlrabi ($2), garlic scapes ($2.5), kale ($2.5), arugula ($4), cilantro ($2). Total value: $26.00

SHARE 2 (June 20): lettuce ($2), turnips ($2.5), beets ($2.5), cucumbers ($4), summer squash ($4), green onions ($2), broccoli ($3), mustard greens ($2), bok choy ($2.5), dill ($2). Total value: $26.50

SHARE 3 (June 27): lettuce ($2), spinach ($3), radishes ($2), cucumbers ($4), summer squash ($4), kale ($2.5), garlic scapes ($2.5), kohlrabi ($2), basil ($2), snow/snap peas ($3). Total value: $27.00

SHARES 4 to 17 (July 4 to October 3): lettuce and, subject to availability, carrots, turnips, beets, cucumbers, tomatoes, summer squash, snow/snap peas, beans, broccoli, cauliflower, garlic, onions, Swiss chard, basil, eggplants, peppers, cherry tomatoes, leeks, melons, tomatillos, celery root, hot peppers and herbs.

SHARE 18 (October 10): lettuce ($2), carrots ($2.5), turnips ($2.5), cucumbers ($4), tomatoes ($4), garlic ($2), leeks ($3), arugula ($2), bell peppers ($3), cilantro ($2). Total value: $27.00

What Goes into Shares at Les Jardins de la Grelinette

The vegetables we grow for our CSA service are chosen based on demand. Over time, we have learned which vegetables our members enjoy receiving frequently and which they prefer to receive less often.

FREQUENT VEGETABLES: tomatoes, lettuce, herbs, cucumbers, carrots, summer squash, peppers, onions.

SECONDARY VEGETABLES: garlic, beets, turnips, radishes, snow/snap peas, beans, broccoli, cauliflower, potatoes, eggplants, Asian greens, arugula, spinach, basil, melons, cherry tomatoes, Swiss chard, kale.

OCCASIONAL VEGETABLES: fennel, hot peppers, tomatillos, chicory, kohlrabi, celery root, celery, winter radishes, winter squash, garlic scapes, sweet corn, Brussels sprouts.

We also use the following principles when putting together the weekly vegetable shares to make our offerings as diverse and enjoyable as possible:

- We include between 8 and 12 different items per share on any given week of the season.

- We include lettuce in every share (or spinach—early and late in the season).

- We include one variety of greens per week (sometimes two or three early in the season) but never give the same greens twice in a row.

- We try to include two root vegetables in every share—carrots as often as possible.

- We try to have fruit vegetables ready as soon as possible to add value to our shares early in the season.

- We include herbs in every share.

We don't expect to please our entire membership with every share, of course, but we are receptive to member feedback and do make an effort to accommodate their preferences. Having said this, it is important not to let member feedback sway your production decisions too much. It is also advisable not to grow too many different kinds of vegetables in your first season. In my opinion, new market gardeners would do well to limit their production to twenty kinds of vegetables for the first few seasons, and start by mastering the more popular ones and those which are the easiest to grow. For the other vegetables, you may consider purchasing some of your offering from other organic growers for your first few seasons. We've always bought potatoes, winter squash, and often watermelons to add to our share. We are transparent about it, and it never bothered any of our CSA partners.

SHARE 19 (October 17): spinach ($3), beets ($2.5), winter radishes ($2.5), cucumbers ($4), kale ($2.5), cauliflower ($3), celery root ($2), onions ($3), broccoli ($3), parsley ($2). Total value: $27.50

SHARE 20 (October 24): spinach ($3), carrots ($2.5), turnips ($2.5), garlic ($4), Chinese cabbage ($4), kohlrabi ($2), leeks ($3), arugula ($2), potatoes ($3), thyme ($2). Total value: $28.00

SHARE 21 (November 1): spinach ($3), carrots ($5), kale ($2.5), onions ($3.5), winter radishes ($2.5), celery root ($2), winter squash ($4), parsley ($2), potatoes ($3). Total value: $27.50

How Much and When to Grow

Once we have decided what to include in our vegetable offering, we need to determine how much to grow and when to grow it. As I began to explain earlier on, we do this by calculating the number of beds needed to generate 220 units per week, as well as the seeding dates required to have the harvest ready as planned. To help us in this task we use a production calculation table like the one on page 141* and go through every seeding, systematically writing down our result on a list like the one presented below. The next step is to combine all the same vegetables together, thereby giving us an overview of how many total beds we will grow of each crop and, in certain cases, how many successions are called for. We then use this information

* Remember that we are working with intensive spacing, which produces higher yields than may be suggested by similar tables based on different cropping systems.

to order seeds for the whole year, while deciding upon cultivars.

Establishing a Crop Calendar

Now that we've precisely determined the number of beds, and the seeding dates for every seeding to be done during the season, the next step is to consolidate this information into a procedure that is easy to understand and implement. We do this by writing down the data for every crop in a monthly desk calendar (using the codes **H** = harvest, **I** = indoor seeding, **DS** = direct seeding, and **T** = transplanting).

At that point we also take the time to systemize other cultivating practices in our calendar. For every planned transplanting or direct seeding,

Planting Dates and Quantities

H = harvest I = indoor seeding DS = direct seeding T = transplanting

Spinach	H: Jun 16	I: Apr 22	T: May 9	2 beds (Tyee)
Radish	H: Jun 16	DS: May 10		1 bed (Raxe)
Kohlrabi	H: Jun 16	I: Apr 6	T: May 3	1 bed (Korridor)
Summer squash	H: Jun 16	I: Apr 26	T: May 16	3 beds (2 – Plato, 1 – Zephyr)
Eggplant	H: Aug	I: Apr 7	T: Jun 1	3 beds (2 – Beatrice, 1 – Nadia)
Spinach	H: Oct 20	I: Aug 1	T: Aug 25	2 beds (Space)
Etc.				

Our crop calendar is crucial for achieving all the planting and upkeep required for our production season. Although there are many ways to organize seeding dates (computerized spreadsheets, specialized software, etc.), we prefer the visual clarity of an at-a-glance paper calendar.

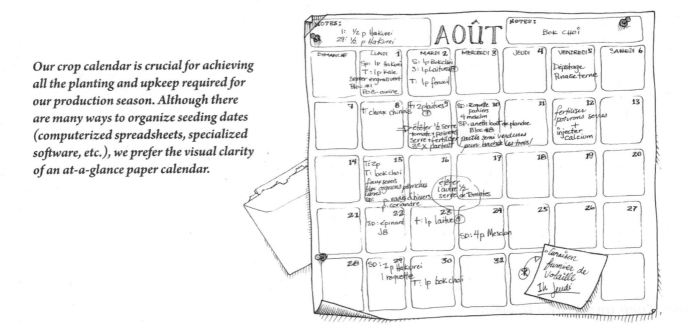

we write the date on which the beds must be prepared, factoring in the time needed beforehand for stale seeding (generally two weeks). We also write in the phytosanitary interventions we want to remember, along with any other information we are liable to forget. For example, if we know that our broccolis will need to be supplemented with boron and molybdenum 10 days after being transplanted, we write that down. If we know that our greenhouse tomatoes need to be fertilized every month, we write that down as well. So on and so forth.

If done properly, a calendar of this sort leaves nothing to chance. It tells us at a glance all the tasks that await us in a given week. At any time during the growing season, our calendar is there to tell us what we should do and when. I cannot overstate how truly vital this tool is to the success of our growing season.

Making a Garden Plan

The final step in our winter crop planning session is to make a garden plan that determines the exact placement of all our yearly seedings. Recall that our farmland is divided into 10 plots of equal size, each one occupied by a botanical family or group of vegetables that follow a precise rotation (See Chapter 6, Fertilizing Organically). When describing how we went about establishing our rotation plan, I mentioned that the consequences of the plan can be somewhat restrictive. Now that we've gone about deciding what to grow and how much we will need of each crop, we also need to make sure that we have enough room to fit everything into the garden, *and* within the spatial constraints imposed by our rotation plan.

What we do is make a map that shows the 16 permanent beds of each of our 10 plots, labeled by family. The task then is to find the best spot for each seeding, taking into account not only the botanical family, but also grouping crops based on specific characteristics that will make maintenance operations more efficient. Crops that might be grouped together are those that go in the ground the same day, those that require floating row cover, or those that we direct seed and irrigate with our sprinklers—which water a defined width, in our case 4 beds. We also want to group crops that will be harvested around the same time, so that succeeding crops can be started following the same routine.

To make this variable clear, we need to know how long each crop will stay in a bed, from the date it is put in the ground to the last harvest date. For each seeding, we therefore write down not only planting dates, but also estimated harvest dates, to which we add a buffer period of 14 days to make sure the crop has time to grow to full maturity. Then we draw a line indicating that the bed is once again free for another crop of vegetables or cover crop. As an example, I have included a rough version of one of our garden plans in the appendices.

As you might imagine, considering all these variables when deciding where to place our crops makes for a lot of shuffling around. Although this task makes our annual planning more complex, it is important because it allows us to use our small area as efficiently as possible. We usually tackle this job while it is still snowing outside, when we have the luxury of time. This saves us from having to worry about crop placement decisions in the summer.

Record Keeping

As I mentioned earlier, once we have developed both our crop calendar and garden plan, we follow though without asking ourselves further questions. This is why it is so important to be really concentrated and focused when doing this work. Developing a crop plan a few days before the seed order deadlines is a really bad idea. Even so, however good a plan we might concoct, it is almost certain that we will discover a mistake in the plan at some point during the season: a miscalculation of the number of days spent in the fields, crop successions too far apart, not enough beds to meet the demand, etc. To prevent these mistakes from happening again the following year, we record all our observations in a loose-leaf binder with a section for each crop. For each cultivar, we note the actual seeding, planting, and harvesting dates along with the yield produced. We also record the number of beds planted, which allows us to adjust our targets for the following year if necessary. At the bottom of each page, we leave a space to note any other useful observations. The sheets in our binder resemble the table on the following page.

Our record keeping system is not complicated or time-consuming, but we've learned that it works best when information is recorded on a regular basis, and so we try to do that. Ultimately, crop planning is all about small details that make a big difference.

Record Sheet Example

Eggplant: 1 row spaced every 18 inches. Fertilization: 5 wheelbarrows of compost and 1.6 gallons of poultry manure.

Date and cultivar	Cell size and number	Date and transplanting location	Date of first harvest	Yield per 100-foot bed
April 7: 3 beds of Beatrice	225 pots (4" × 4")	May 30 in the garden	July 5–17	
April 7: 1 bed of Nadia	75 pots (4" × 4")	May 30 in the garden	July 5–25	

Intervention July 5: pyrethrum to reduce tarnished plant bug population.

Intervention August 30: pyrethrum to reduce tarnished plant bug population.

Note: Beatrice produces well all summer despite presence of tarnished plant bug.

We take notes of anything that goes wrong as soon as we realize it. We know from experience that otherwise we will forget.

Production Calculation

Vegetable	Days to maturity*	Yield/100-foot bed**	Notes
Arugula	35	200 bunches	
Asian greens	60	300 units	
Basil	60	150 units/week	Assume 1 bunch (0.4 oz.) per plant every 2 weeks once production is underway.
Bean	55	65 lb./week	Assume 130 lb. total over 2 weeks of good production.
Beet	60	160 bunches	
Broccoli	75	120 heads	
Bulk spinach	40	75 pounds	Assume 35 lb. at the first harvest and 40 lb. between the second and third harvests.
Cabbage	80	150 units	
Carrot	55	180 bunches	
Cauliflower	75	130 heads	Allow many more days in the garden for certain cultivars.
Celery root	140	300 units	
Eggplant	100	65 units/week	Assume about 1 fruit per plant per week once production is underway.
Fennel	80	400 units	
Garlic	N/A	600 units	
Green onion	75	350 units	
Greenhouse cucumber	50	115 units/week	Assume 1.75 fruits per plant per week.
Ground cherry	110	Undetermined	2 beds are enough for our annual needs.
Kohlrabi	60	420 units	Allow many more days in the garden for certain cultivars.
Lettuce	50	250 units	
Melon	80	100 units or less	Assume 1.25 fruits per plant.
Onion	120	400 lb.	
Pepper	120	120 units/week	Assume about 1 fruit per plant per week once production is underway.
Radish	30	300 bunches	
Snow/snap peas	55	25 lb./week or less	Assume 75 lb. total spread out over 3 weeks of good production.
Summer squash	50	100 units/week	Assume 2 fruits per plant per week.
Swiss chard and kale	60	150 units/week	Assume 1 bunch per 2 plants every 2 weeks.
Tomato	120	150 lb./week	Assume 3 fruits per plant per week once production is underway.
Turnip	40	200 bunches	
Summer leek	120	175 units	These are sold in bunches of 3 or 4.

* "Days to maturity" refers to the number of days between the seeding and the first harvest. This expression is not synonymous with "days in the garden" (which I mention in the crop notes), as the figure for days to maturity includes the time spent in cell flats. For vegetables transplanted early in the spring or late in the fall, the days to maturity must be adjusted for slower plant growth.

** The weekly yields are approximate and take account of our intensive spacing on a bed 30 inches wide and 100 feet long. You will have to modify these numbers in proportion to the dimensions of your beds.

Conclusion
Farming for Ecology, Community, and Lifestyle

Like multitudes of people all over the world, we are seeking a good life—a simple, balanced, satisfying life style. Like them, our aim is to lend a hand in shaping the planet into a homelike living place for successive generations of human beings and for the many other life forms domiciled in and on Mother Earth, her lands, and water.

—Scott and Helen Nearing, *Living the Good Life*, 1954

It is with gratitude that I quote Scott and Helen Nearing in this conclusion to *The Market Gardener*. Their book *Living the Good Life: How to Live Sanely and Simply in a Troubled World* has provided, and continues to provide, inspiration to scores of young people seeking a simple life based on the principles of sustainability, ecological agriculture, and self-reliance. The Nearings' story about sustainable living, although written more than forty years ago, is still relevant today, as it provides an effective antidote to mainstream consumerism. It has been one of my favorite books for many years, especially at the time when my wife and I were driven by the desire to pursue meaningful careers, and to reconnect with nature in our everyday lives. Considering how much of our time on Earth is spent working, why couldn't

our everyday labor serve the higher ideals we were setting out to live by? In answer to this, the Nearings provided us with a great example.

For us, farming came to be the answer to the alternative lifestyle that we sought, and I feel privileged to have found such a satisfying calling so early in life. Our profession has allowed us to homestead, built our own house, and raise our family in the countryside, all while being immersed within the rhythms of the changing seasons. Don't get me wrong, farming is hard work, and everything we sell we've had to grow using our own sweat and labor. But when working for a purpose, hard work is satisfying. Looking back, I also realize how farming has connected us to something much bigger than the farm and the lifestyle we set out to create for ourselves.

For one thing, it allows us to participate in society without being completely embedded in the globalized economy. The vegetables we sell are grown from seed, using minimal amounts of fossil fuel. The tools we use all come from small companies, and none of the fundamental inputs required for growing our crops come from an industrialized process. By selling directly to customers and bypassing distributors, we ensure that both we and our clients are generously rewarded in this relationship. Effectively, we are producing a commodity limited to a local and equitable chain of production. To me, this is pretty awesome.

Feeding people locally not only helps our community become more resilient, it also allows us to engage with the community in a special way. Every week at the market, we meet excited and appreciative clients, who are eager to learn about the growing process, and equally eager to cook up the week's harvest. Many of our CSA partners have told us how we are included in their prayers of gratitude before every meal. Others have remarked that, beyond being food providers, we also serve as a link to the natural world for them. For all of these customers, our produce is not like any other commodity, but has a special place in their lives. All the positive feedback we receive is perhaps the most rewarding aspect of our difficult job.

What I have observed is that the local organic movement is not only creating demand and recognition for our craftsmanship, but is also placing us farmers in a leading social and political role. Farmer colleague, Will Allen, perhaps best captured the scope of this movement with the title of his book *The Good Food Revolution*. Although we may not always be fully aware of it, our line of work is demonstrating how to organize our lives without the globalized economy, making us less dependent on it in the process. Local agriculture does have the power to transform society, and I believe that ultimately it will.

What is so great about all of this is that farmers and consumers are not alone in this paradigm shift. Non-governmental organizations, food activists, community organizers, teachers, health-care practitioners, concerned politicians, and other engaged citizens all participate in this struggle to build a better world, with agriculture as one of its pillars. The amazing work these people do, whether actively lobbying for political reform, supporting direct sales mechanisms, or serving a delicious locally grown organic meal, all help to prevent farmer exploitation and contribute to a higher food consciousness. These people have my deepest respect, and I hope that their energy translates into bringing more and more people to our table. But what this movement needs is, above all, for more of its supporters to put on their rubber boots and take up growing food.

Unfortunately, many young people are easily turned off by the trade, not because of the agrarian lifestyle, but rather by the economic implications of owning and running a farm. Starting an organic vegetable farm often requires heavy mechanized machinery, a substantial land base, labor management, and the acquisition and maintenance of costly infrastructure. Looking from the outside in, purchasing or starting such an enterprise looks inaccessible. I know this feeling, because I felt this way when I got interested in farming.

But growing vegetables for market and making a living at it does not need to be done that way. You can enter agriculture professionally by replacing pricey equipment and infrastructure with a different skill set that is based on appropriate technology and innovative growing practices. My hope is that our experience and the information shared in this handbook will help aspiring farmers understand how this can be achieved.

One of the most encouraging things I see is the number of educated, politically aware, and enthusiastic young people heartily interested in learning the art of growing food sustainably. Soon enough, our group of like-minded people will form a powerful critical mass, and when it does, we will be a force to be reckoned with.

With the demise of cheap oil comes the evolution of artisanal, resilient, biological agriculture. This evolutionary step is not far away—it's just around the corner. It's up to us to reinvent the noble profession of farming. We have not only the choice to do things differently, but the means as well.

— Jean-Martin Fortier
Saint-Armand, Québec
October 2013

Appendix 1
Crop Notes

In this handbook I have shared information about our growing practices from a general point of view. This was essential to give the reader an overall understanding of the intensive cropping system we have developed on our farm. But diversified market gardening also requires technical know-how for each crop to be grown. Every vegetable has its own set of characteristics, each one requiring particular attention and care. This appendix contains notes on how we grow our favorite crops at Les Jardins de la Grelinette.

Legend

 COLD HARDY

 DIRECT SEEDED

 TRANSPLANTED

 HIGHLY PROFITABLE

 EASY TO GROW

 BENEFITS FROM BROADFORKING

Notes

- The recommended densities are optimized for beds 30 inches wide.
- To get an idea of the doses of fertilizer amendments used, refer to the fertilization plan on page 56.
- The seeding dates listed are not set in stone but rather vary from year to year. I present them here as a reference when developing a first crop plan. Also note that our farm is located in southern Quebec, a zone 5 area in the plant hardiness zone map.

INTENSIVE SPACING: 5 rows 6 inches apart, spaced every 1¼ inches
PREFERRED CULTIVARS: Arugula, Astro (winter)
FERTILIZATION: Light feeder
DAYS IN THE GARDEN: ± 45 days, including two cuttings
NUMBER OF SEEDINGS: 4 (May 10, May 20, August 25, September 1)

Arugula *(Brassicaceae)*

Arugula is no longer a foreign vegetable to Quebecers; most have heard of it and enjoy its unique peppery taste. It is one of our most popular greens at market, grows quickly, and is frost-hardy—surviving temperatures as low as 23°F/−5°C. We can always count on having arugula for early and late harvests, but we prefer to avoid it during the summer since its taste can become a bit sharp at that season. Nevertheless, the demand for arugula never ceases, and it would certainly be possible to grow it all summer long in a cool, dark location in the garden or under a shade cloth.

Arugula is particularly susceptible to damage by the flea beetle, which pierces holes in the leaves and makes them much less attractive to customers. We keep ours covered with anti-insect netting or row cover at all times. We sell arugula in bulk (washed and dried like mesclun mix) and in bunches with its roots on. To maintain the highest quality, we never do more than two cuttings of arugula on the same bed.

Beans *(Fabaceae)*

Ah, beans! So popular, so much work. Customers just love fresh beans, but many market farmers choose not to grow them because of the difficulty of getting them picked economically. While picking beans is labor-intensive, the demand and scarcity have always led us to grow them.

Because beans do not tolerate frost, our first crops are never early. We aim to have beans ready by mid-season, when our snow and snap peas are just running out. We try to keep a steady supply of beans at market and like to include them a few times per season in our CSA shares. We accomplish this by harvesting a bed for only two weeks before moving on to another succession.

We therefore favor bush beans (as opposed to pole beans) as these are more productive over the short term. Since beans *absolutely* need to be picked every two to four days (otherwise they grow too big, losing their flavor and tenderness), we are extra careful not to stagger our successions so that they overlap one another. We do this by simultaneously seeding two cultivars with different days to maturity.

The EarthWay seeder makes bean seeding easy. In fact, I know a number of growers who keep this seeder only for this crop. Beans need warm soil early in their growth, thus a row cover is a must for early springtime seedings. They grow well without extra fertilizer and have no problematic pests in our garden. Rust and mildew are common fungal diseases for this crop, but our rapid turnover from an older crop to a newer one never allows these pathogens to develop. Apart from harvesting, the only upkeep required for bush beans is hoeing. Bean plants grow rapidly, which quickly makes hoeing between the rows impossible. Keeping weeds at bay therefore requires that the crop be hoed regularly in the beginning stages.

The ideal bean is fine and not stringy. To ensure quality beans, we pick them while they are only as wide as a pencil at the most, and definitely before the seeds become bulbous in the pod. Freshly picked beans should not be washed (they can otherwise form a mold) and must be cooled in the cold room in sealed bins. In ideal conditions, beans keep for up to a week after being picked.

INTENSIVE SPACING: 2 rows 14 inches apart, spaced every 4 inches

PREFERRED CULTIVARS: Provider (tasty, early, and reliable), Maxibel (filet bean), Jade (summer), Rodcor (yellow), Edamame (soybean)

FERTILIZATION: None

DAYS IN THE GARDEN: ± 70 days, including two or three weeks of harvest

NUMBER OF SEEDINGS: 5 (May 23, June 6, June 21, July 4, July 20)

Beets (Chenopodiaceae)

When talk turns to beets, many of our customers and partners report having rediscovered the joys of this garden "classic." The beet's days as a canned vegetable are hopefully behind us, as this delicious vegetable is making a comeback. We now sell just as many bunches of beets as we do of carrots.

Growing beets is not complicated, but one special feature of this crop is that what appears to be its seeds are actually clusters of three or four smaller seeds fused together. Because of this, beets are one of the crops that we must thin to the desired density when direct-seeded. Seed developers have hybridized monogerm cultivars to avoid this problem, but there are few to choose from. This is why we prefer to transplant this crop. We start them in cell flats of 128 and set them out in the garden at optimal spacing. The resulting high yield of nice round beets is well worth the trouble.

Beets also have the distinction of being one of the few pest-free crops. Scab can be a common bacterial disease, especially in crop rotations that include potatoes, but since we don't grow potatoes we've never faced this problem. Late in our season, we often see Cercospora leaf spot, a fungus that creates brown spots and holes in the leaves. But it doesn't bother our customers, who are more interested in the roots than the tops.

Beets are well-adapted to variations in temperature, which means they can be grown all through the season. There is some latitude with respect to harvesting them (they stay tasty even if kept in the soil for a long time), but we prefer picking them when they are 2 to 3 inches in width, since this is what our customers prefer. When bunching, we clean off the dead leaves and put 3 to 5 beets of the same size (so that they cook more evenly) per bunch. If we have a surplus of unsold beets at the market, we simply cut off the leaves and put the roots into storage, to be sold later at a lower price to customers who are looking for beets to preserve.

INTENSIVE SPACING: 3 rows 10 inches apart, spaced every 3 inches

PREFERRED CULTIVARS: Early Wonder (early planting), Moneta (direct seeding), Red Ace (keeps well in the soil), Touchstone Gold (yellow), Chiogga (simply beautiful)

FERTILIZATION: Light feeder

DAYS IN THE GARDEN: 50 to 60 days, including two or three weeks of harvest

NUMBER OF SEEDINGS: 6 (March 28, April 18, April 20, May 10, June 6, June 30); started as transplants

Broccoli (Brassicaceae)

Broccoli is a popular vegetable that is pleasing to grow, but it does have its challenges. It's a heavy feeder that requires a lot of nitrogen and potash. This is one crop where a supplementary dosage of organic fertilizer, in conjunction with compost, is essential. Preceding the crop by a leguminous cover crop can also make for dark green broccoli heads, which is what you want.

Broccoli grows best in cool weather, but we try to grow it all through the season to meet the high demand. We do two succession plantings in the spring, one in the summer, and another in the fall. The summer crop is the most challenging, as broccoli tends to flower quickly in hot weather. Certain cultivars are resistant to bolting and give us satisfactory results.

For optimum growth, broccoli requires a number of systemic interventions. The crop must receive regular supplements of boron and molybdenum, as most soils in the Northeast are deficient in these two micronutrients. Molybdenum deficiency manifests as fragile leaves which break easily when handled. Boron deficiency shows up as ribbed, crimped leaves. In both cases, one or two sprays of boron mixed with molybdenum during the plant's growth will keep these deficiencies in check. In this mix, we also add a preventive application of *Btk* (*Bacillus thuringiensis* variety *kurstaki*), which kills off caterpillars that are often feeding on the plants. This intervention is planned 15 days after transplanting the crops and 10 days before the broccoli matures.

Unfortunately, a new insect pest has appeared on all *brassicas* in the last few seasons: the Swede midge. This insect inhibits broccoli growth—and only a few are needed in a garden to make the whole harvest unsellable. We have encountered this pest only once on our farm, but the damage was so severe that we now protect all of our *Brassicaceae* with anti-insect netting held up by hoops. The netting we use for the Swede midge also serves as protection from the cabbage maggot, flea beetle, and cabbageworm—so in a way this new pest may have had a silver lining.

Timing the harvest is key to a successful broccoli crop. The heads mature very quickly (we have seen them double in size within a 24-hour period!), and in order to harvest broccoli at its peak (i.e., when it is well developed, thick, and dense), the harvest must be done over several consecutive days. Determining when the head is fully mature is not so obvious. You have to regularly inspect and measure the flower cluster size, harvesting just before the small individual buds begin to flower. At that point, the broccoli heads are generally 4 to 8 inches in diameter. If you notice the flower clusters turning yellow, or that flowers are beginning to separate or open, you should cut the head off right away, as it is starting to bolt. To extend the harvest season, you can also encourage the growth of secondary shoots by cutting the head of the broccoli high on the stalk during harvesting. This will cause the base to put out side shoots that will eventually yield broccoli florets. We often bunch these together and sell them for the same price as the main heads.

We harvest broccoli with a machete-style harvest knife, aiming to have a 5–6 inch stalk on each head. Once harvested, broccoli must be cooled of its field heat with cold water and refrigerated immediately; otherwise it rapidly loses its freshness. Broccoli can keep crisp for more than a week if kept in the cold room at a low temperature (less than 39°F/2°C).

INTENSIVE SPACING: 2 rows 14 inches apart, spaced every 18 inches

PREFERRED CULTIVARS: Packman (spring, very early), Gypsy (summer), Windsor (fall)

FERTILIZATION: Heavy feeder

DAYS IN THE GARDEN: 65 days after transplanting

NUMBER OF SEEDINGS: 4 (April 17, April 24, May 28, June 25)

Carrot (*Umbelliferae*)

Our garden carrots are our most effective marketing tool. Whenever we meet naysayers who are not convinced about the virtues (and price) of organic produce, we don't argue; we simply hand them a carrot. One bite is usually enough to convince anyone of the difference between our vegetables and those from the supermarkets. We grow ours with flavor in mind, and this is why we mostly cultivate Nantes carrots, which are the sweetest and tastiest of all. We also sell our carrots in bunches with their tops, thereby indicating the freshness of the product. A market gardener can always get a premium price for a bunch of freshly picked juicy carrots.

Since carrots are both popular and cold-hardy, they are an excellent vegetable to force in the spring. Our first seeding is planted in the hoophouse using the Six-Row Seeder at a density of 12 rows per bed. We then use pelleted seeds to produce early baby carrots that sell like hotcakes at our first market. Our second early seeding is done directly in the garden, but under the cover of a caterpillar tunnel and a row cover. At that point, we aim to grow bigger sized carrots and space the seeding to 5 rows on the bed. From what we've observed, carrots need about 1½ square inches to grow to a size of 6 to 7 inches, which is what we aim for.

We time our final carrot seeding so that the crop can be harvested from the field right until our very last shares in October. When the nights begin to get frosty, we cover our last carrot beds with row covers, sometime two. Carrots store well in the ground as long as the temperature doesn't sink below 20°F/−7°C. The cold nights actually make the carrots taste much sweeter, much to the delight of our CSA members.

Carrots like loose, well-aerated soil that's been worked quite deeply with the broadfork. This crop is not a heavy feeder, and fertilizing with compost once every two years seems to be enough to meet its needs. Carrots don't appreciate high-nitrogen fertilizer or excessive fresh manure, demonstrating this aversion by forming hairy roots. However, we do apply supplementary fertilizers in our first seedings when the soil is still too cold to provide sufficient nitrogen for good leaf growth.

The biggest challenges with carrots are seeding and weeding. Since carrots take a long time to germinate (usually 8 to 15 days), it is important that the soil surface be kept very moist so that it does not harden or dry out by the time the fragile plants emerge. We accomplish this by installing micro-sprinklers to water as needed. We also lay floating row cover over the soil to help the plants come up faster. The best weed-control strategy we have come across is pre-emergence flame weeding (see Chapter 9). This technique saves us untold hours of weeding on our knees. A flame weeder, even if used only for carrots, is definitely a worthwhile investment.

Carrots also have some insect challenges. Two pests, the carrot rust fly and the carrot weevil, are responsible for the brownish scarring seen on affected carrots. The rust fly causes the most problems for us, but we manage to fend it off by covering the crop with anti-insect netting in mid-August, during the fly's egg-laying period. Weevil damage has been minimal on our farm; our solution is simply to pick the affected carrots out and sell them to our customers as juicing carrots. During seasons that are unusually wet, a number of diseases sometimes affect the leaves, but this usually happens when the carrots are mature. Our solution is simply to cut off the upper half of the tops.

Carrots develop color and taste at the same time. This means they can be harvested as soon as they look good to eat. At harvest time, we pull a few to check their size, loosen the soil with a fork, then gently pull them out of the ground and bring them to storage to be bunched fresh. Watering right before harvest also makes pulling easier. One should not wait too long to harvest carrots (especially summer crops) as they will lose flavor and develop a different texture if left in the soil for too long. If carrots are split, they should have been harvested earlier.

We generally put 8 to 12 carrots in a bunch. If we spot any carrots that have been speared by the fork or that are otherwise worse for wear due to our gravelly soil, these go in with the juicing carrots. We cut the tops off our unsold bunches and store leftover carrots in the cold room, saving them to make giveaway bags for our CSA partners at the last delivery. Stored carrots will easily keep in the cold room for up to 6 months.

INTENSIVE SPACING: 5 rows 6 inches apart, spaced every 1¼ inches

PREFERRED CULTIVARS: Nelson (pelleted), Yaya (summer), Purple Haze (purple), Napoli (last seeding)

FERTILIZATION: Light feeder

DAYS IN THE GARDEN: ± 85 days, including two or three weeks of harvest

NUMBER OF SEEDINGS: 8 (April 10, April 25, May 4, May 25, June 8, June 23, July 5, July 25)

Cauliflower (Brassicaceae)

Cauliflower is grown similarly to broccoli. Crop requirements, plant protection measures, and harvest instructions for this plant can all be found

in our notes on broccoli. However, cauliflower has some unique characteristics that present extra growing challenges. This crop is definitely high maintenance, and it might require a few trials before growing it right.

For one thing, cauliflower is not as adaptable to changes in the weather as broccoli. It tolerates only light frost and requires consistently cool temperatures in the 60°F/15°C range. Any stress tips the balance toward premature heading, or "buttoning," when the plant makes tiny buttons rather than forming one nice white head. Success with this crop is highly dependent on temperatures throughout the season, which basically means that bountiful harvests are never guaranteed.

To ensure that we grow in the best conditions, we do two seedings of cauliflower: a very early one in the spring (which we cover with row cover during much of its growth) and a second one late in the summer (so that the heads reach maturity before the cold weather sets in). In both cases, we put anti-insect netting over the crop to protect it from Swede midge.

Customers who look for cauliflowers want them to be white. Yellow ones, green ones, or any other colors simply don't sell as well. To grow cauliflower with a nice white head, the crown must be kept shaded for the last sprint of its growth. This is called blanching, and we do this by covering the head with the side leaves of the plant as soon as the head begins to form, i.e., about 5 to 10 days before harvest. The leaves can be held together with an elastic band or simply broken in such a way that they cover the head. We prefer the latter technique as it allows us to quickly check whether the vegetables are ready to harvest.

As with broccoli, you must wait for the head to become firm and compact to harvest it at its peak; this happens just before the plant flowers, usually when the heads have grown 6 to 8 inches in diameter. If the heads are too small but have started to open up, they will not improve and should be harvested. Freshly picked cauliflower must be cooled quickly and stored in the cold room before being sold.

INTENSIVE SPACING: 2 rows 14 inches apart, spaced every 18 inches

PREFERRED CULTIVARS: Minuteman (spring), Bishop (fall)

FERTILIZATION: Heavy feeder

DAYS IN THE GARDEN: ± 80 days after transplanting

NUMBER OF SEEDINGS: 2 (April 24, June 15)

Chicory *(Asteraceae)*

In Quebec, chicory, or curly endive, is not a popular vegetable as most people tend not to enjoy bitter flavors. The same is true of radicchio, endive, and escarole, crops that are even less well-known than chicory. But tastes are changing, and more of our customers are asking us for these traditional European gourmet greens.

We grow chicory to add texture and volume to our mesclun mix. Of all our ingredients, chicory is the easiest and most reliable green of our mix. It is cold-hardy, yet grows well in hot weather, has no predators, and allows many cuts from the same plant. We grow it from transplant and lay out the garden at spacings that allow us to grow mature-sized heads. The part we want is the heart of the plant, which is composed of fringed whitened leafy shoots that are extra tender. At harvest, we cut the whole head off (to ensure good regrowth) but slice the chicory in half leaving out the outer leaves and the tops. To make a neat cut, we use a long-bladed knife (our favorite is the Opinel from France) and use the side of the harvest bin as a cutting board.

Once cut, the plants grow back new leaves, always leaving the inside ones young and tender. The bigger the plant gets, the more blanched and elongated the leafy shoots will get. To further enhance this effect, you can pull the outer leaves together with a rubber band, leaving it to blanche for a week or two. By caring for the crop this way, you can get just as many cuts as you want from the same bed, which is a good tradeoff for the extra effort put into tying up the plants. We process chicory leaves the same as we do for our other meslcun mix ingredients.

INTENSIVE SPACING: 4 rows 6 inches apart, spaced every 6 inches

PREFERRED CULTIVARS: Tres fine (curly), Rhodos

FERTILIZATION: Light feeder

DAYS IN THE GARDEN: ± 75 days for three cuttings
NUMBER OF SEEDINGS: 2 (May 11, July 11)

Cucumber (Cucurbitaceae)

For some years we tried to grow field cucumbers, but without much success. Now we grow them exclusively inside our hoophouses. The plastic cover greatly reduces common disease vectors that are caused by the splattering of the soil onto the plants. The structure of the hoophouse also makes it convenient to trellis the plants to grow vertically, which in turn offers the greatest return per square foot. Three 100-foot beds of trellised cucumbers supply all of our CSA and farmers' market needs. Grown in such a way, cucumbers are second only to tomatoes in terms of profitability.

There are many interesting cultivars of greenhouse cucumbers to choose from, but we have two favorites: a long English cucumber and a Lebanese short cucumber. Both cultivars are seedless and resistant to many diseases. Neither requires pollinators to be able to fruit, which is ideal since our hoophouses are closed off by insect netting.

We start our plants in cell flats, aiming to have cucumbers at our first markets. They usually stay in their cells for about 15 days before being transplanted, but for our first seeding we do things differently. In early May the soil in our hoophouse is still cold, so we pot up the fragile seedlings into large 6-inch containers, enabling them to enjoy another two weeks of optimal growing temperature in the plant nursery. In the meantime, we work on warming the soil by laying old greenhouse plastic over the beds where the cucumbers are set to grow. Two weeks of solarisation is sufficient to bring up the soil temperature to the 64°F/18 °C that is needed for optimal plant growth.

Cucumber seedlings are fragile, and we try our best to handle cucumber plants as gently as possible when potting up and transplanting. Disturbed roots at this stage of growth might cause the plant to be more susceptible to various diseases. We are also extra careful not to bury the neck of the transplants in order to prevent neck rot. Once they are properly planted, we tie the cucumber plant to a vertical wire suspended from the steel brace that runs the length of our hoophouses. Then begins the rigorous work of keeping shoots and fruit pruned, maintaining a proper balance between the vegetative growth and fruit load to maximize production.

For the first few weeks, until the plant is about 2 feet tall, we remove all the flowers and/or small cucumbers that appear. This allows the plants to establish better root systems, which in turn allows them to produce more fruit in the long haul. When the plant has produced six leaves (nodes), we continue to prune leaving only one fruit for every two nodes (for the Lebanese type, we leave two fruits per node). Pruning is important because if too many fruits are allowed to form at any one time, a large proportion will abort, causing the cucumbers to be malformed or poorly coloured. We keep pruning in this way, making sure to remove all laterals, until the stem reaches the height of the steel brace. At that point, we let one shoot grow in addition to the main stem so as to have two stems per plant trained over the wire. The main stem then comes down, and we keep pruning it, now leaving one fruit per node (for the Lebanese, we stop pruning it). When the main stem develops its sixth node coming down the brace, we cut the head off to allow the plant to focus its energy on the new stem, which is now treated as a main stem.

This pruning method is known as the umbrella system. Theoretically, a renewal system such as this makes it possible for a single plant to produce fruit for a whole season, but we often don't make it that far. After seven or eight weeks of harvest, many of our plants die of bacterial wilt, a virulent disease transmitted by cucumber beetles who manage to get inside our hoophouses, even though we've sealed them off with insect netting. It only takes a couple of sneaky insects to cause damage, and since we do not want to use repeated applications of insecticide just to kill a few insects, we concede defeat against the intruders and start again with a fresh cucumber crop. This means that we plan two seedings per season, one replacing the other in each of our two hoophouses. Between the two cucumber plantings, we use the available space to grow a short green manure cycle.

Most years we encounter other insect pests

such as thrips and spider mites in our cucumber hoophouse. We deal with these by introducing predatory mites that feed on the larvae of the two pests. To successfully multiply these predatory mites requires a moist environment. We've thus equipped both of our hoophouses with a simple misting system regulated by a water timer. This biological pest control increases our production costs, but the results are worth the investment.

As we do with our tomatoes, we also fertilize our cucumbers with a mix of vermicompost and chicken manure hoed in at the bottom of the plants. We side dress the crop in two rounds: one when the soil is prepared and another four weeks after transplanting. The latter round includes a dose of potassium sulfate for good fruit development. We water the crop by drip irrigation and cover the soil with tarps, as we do for tomatoes. Under these conditions, each plant can produce two to three cucumbers per week (twice as much for the smaller Lebanese) depending on the outside weather.

The ideal length for harvest is 8–14 inches for English cucumbers and 6 inches for Lebanese cucumbers, which means picking them every two days, or every three in cloudy weather. Immediately after being picked, the cucumbers are immersed in a cold water bath and stored in the cold room in closed bins that drain. They stay crisp for about a week in these conditions. Cucumbers will keep for even longer than this if individually wrapped in plastic film.

INTENSIVE SPACING: 1 row, spaced every 18 inches
PREFERRED CULTIVARS: Sweet Success (English), Jawell (Lebanese)
FERTILIZATION: Heavy feeder

DAYS IN THE GARDEN: 55 to 75 days between plantings
NUMBER OF SEEDINGS: 2 (March 25, July 10)

Eggplant (Solanaceae)

Eggplants, or aubergines as we call them in French, are becoming increasingly popular at market. The diverse shapes and colors of the various cultivars on display make them hard to resist. Eggplants belong to the Solanaceae family and are grown in a manner similar to fellow nightshades tomatoes and peppers. They require a hefty dose of fertilizer and grow fastest in hot conditions with constant irrigation. They are perfect candidates to be grown under plastic mulch, which enhances these effects.

We grow our eggplants under landscape fabric in which we've burned holes according to our favored spacing. We usually grow three or four beds of eggplants per season and use a fabric that covers the whole area, thereby keeping weeds to a bare minimum. Like all of our other vegetables grown under plastic mulch, we water the crop by installing drip tapes set to a water clock.

The main challenge with eggplants is insect pests, most notably the Colorado potato beetle, which ferociously feeds on the leaves of young seedlings. To keep our plants out of harm's way, we install row covers, supported by hoops, immediately after transplanting. The row cover also serves as a beneficial windbreak to the eggplant transplants, which are particularly vulnerable to wind damage. The combination of mulch and row covers allows for optimal growth in the first few weeks after transplant. We remove the row cover when the plants begin to flower, which coincides with the time when potato beetles are less intense. From then on, we scout our rows every week and handpick the larvae of any infestations we come across.

The tarnished plant bug is another insect we watch out for. This insect is hard to see and will cause beautiful plants to not bear any fruit. When we notice that flower buds have fallen to the ground, the tarnished plant bug is usually to blame. Field scouting for this insect involves shaking selected plants (15–20) sampled across our rows over a piece of white cardboard. Doing so will dislodge the nymphs, and if we detect them on more than one fifth of the plants examined, then we intervene with a natural insecticide.

Eggplants can be harvested at any stage of growth, as this only affects the firmness of the fruit. However, we've observed that people tend not to go for large eggplants, so we tend to pick ours when they are small to medium-sized. We also choose cultivars that are on the small side. Eggplants can be harvested in the middle of the day with no problems, but they should then be cooled quickly so that they stay firm until they are sold.

INTENSIVE SPACING: 1 row, spaced every 18 inches
PREFERRED CULTIVARS: Beatrice (round), Millionaire (Asian type, very early), Nadia (classic thick fruit), Fairy Tale (purple and white)
FERTILIZATION: Heavy feeder
DAYS IN THE GARDEN: Entire season
NUMBER OF SEEDINGS: 1 seeding, transplanted after the last spring frost

Garlic (Liliaceae)

Garlic is a great storage crop for the market garden. It's popular, high-yielding, and well-adapted to Quebec's cold climate. The fact that most of the garlic sold in supermarkets comes from China also provides excellent marketing opportunities. Locally grown, high-quality, gourmet garlic will bring a generous price per pound. This being said, growing great garlic is not so easy. Following appropriate procedures for planting, harvest, and storage is critical.

We grow hardneck garlic on our farm because it is more flavorful and generally stores longer than softneck varieties. It is planted late in the fall and harvested the following summer, just in time for our big markets in August and September. We sell it by the bulb throughout the season, but many of our customers want some as winter stock. We thus need to grow a lot of it and devote an entire plot in our crop rotation solely to this crop. Growing more than a thousand garlic plants is no small task, and it must be approached in a highly organized manner.

Every year we invite our CSA partners, friends, and family to a "planting party" to help with this task. Prior to the party, we separate the cloves from the bulbs, keeping only the firmest, biggest, and healthiest-looking bulbs. We prepare our beds by adding compost, broadforking, and working the soil at about 2 inches deep so that it is loose and easy to work into. During the party, we instruct our helpers to plant the cloves pointy side up, a little over an inch below the soil surface, into holes we've punched in the ground using a marking dibbler. Next, we cover the garlic beds with four to six inches of straw mulch. The straw will keep the soil from freezing too quickly, allowing the seed cloves to develop a good root system before winter sets in.

In early spring, when the tip of the garlic leaves start to emerge, we remove some of the straw to allow the soil to warm more quickly. This also prevents problems that might arise from excessive

moisture. In clay soils, and/or beds dominated by weeds, it's wise to remove the mulch completely to allow for frequent hoeing. Garlic does not grow well with weeds nearby, and keeping the crop weed-free is one way to help ensure large bulbs.

By the time mid-June rolls around, the garlic plants will have developed flower-bearing stems known as scapes, which we remove to ensure that the plant is putting its energy into developing a larger bulb size. We harvest scapes two or three times a week for about three weeks in June, and sell them at our first markets when adding diversity to the limited selection of available crops is welcome. By the beginning of July, we begin harvesting fresh (half-cured) garlic bulbs, which we sell individually with the whole stem attached. Then, in the middle of the month, we harvest our entire garlic crop.

Knowing when to harvest is tricky. If harvested too early, there will be fewer layers of bulb wrapper, offering less protection for handling and storage. If harvested too late, the bulbs will often split open. We begin our harvest when about 30% of the plant leaves start to die off, indicating that the plants have cut back on the nutrients and moisture supplied to the leaves. At that stage there are five or six green leaves remaining while the rest of the leaves are yellow and dry. Then comes the critical stage of pulling plants and curing them for storage. How this step is followed will determine the garlic's shelf life.

After experimenting with several techniques, we settled on a procedure that involves cleaning the bulbs in the garden immediately after picking them: we pull the plants by hand and immediately peel off the first leaf (which comes off more easily than when they are cured) to remove any soil from the bulb. We then lay the plant on black geotextile, leaving it to dry in the sun for a few hours. Toward the end of the day, we trim the roots off the bulbs (this prevents the bulbs from absorbing moisture, which can disrupt the curing process) and bring all the garlic bulbs (still with their long necks) inside for storage. The plants are laid out again, this time on stacked seeding tables that are well ventilated by several industrial fans. In these conditions, the garlic will fully cure within three weeks. The final step is to cut the necks to a length of half an inch and put the bulbs into mesh bags. Under good home storage conditions, a solid, well-cured, well-wrapped garlic bulb will keep 6 to 8 months or longer.

As far as phytosanitary problems are concerned, well-fertilized garlic should not be susceptible to any troublesome pests other than the leek moth, which causes minor damage in its first round of egg laying. The most common problems with garlic come from fungal and viral diseases, which cause the bulbs to rot. There are three plausible explanations for rotten bulbs: the soil was too moist when the bulb matured (because of overly thick mulch or poor drainage), the bulbs were not cured properly after harvest, or a virus contaminated the crop via sick seed cloves. The latter is the most common explanation, which is why it is important to plant healthy, disease-free garlic. If you observe any rot in any part of your harvest, it's best not to risk saving these cloves for planting. You're much better off buying new seed garlic from a gardener who specializes in the crop, even if this is a more expensive option. Also, always inspect the seed garlic before purchase;

shipment by mail should be avoided unless the supplier is willing to send you a representative sample of the seed.

INTENSIVE SPACING: 3 rows 10 inches apart, spaced every 6 inches

PREFERRED CULTIVAR: Music (great looking and tasting)

FERTILIZATION: Heavy feeder

DAYS IN THE FIELD: 75 to 90 days starting in May

NUMBER OF SEEDINGS: 1 (Oct. 10)

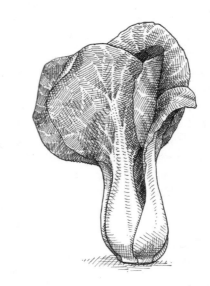

Greens (Asian) (Brassicaceae)

Since we began market gardening, we have been giving CSA members and market customers their first taste of a wide variety of Asian greens. These exciting greens are easy and fast to grow in a cold

climate, so we constantly encourage consumption, hoping to increase their popularity.

Greens like bok choy and Chinese cabbage are an important part of our crop plan as they allow us to diversify our offering in the first few and last markets of the year when fewer vegetables are available. We start them early in our indoor plant nursery and bring them out in the gardens covered with row cover, even when temperatures can still drop below freezing. Asian greens rarely catch a disease, but flea beetles and cabbage caterpillars do love to feast on their tender leaves. Floating row covers in the spring and anti-insect netting in the summer help keep these pests under control. Slugs can also be a problem, and we control them by hand or by using iron phosphate pellets. Apart from these interventions, no special attention is required.

Like all members of the *Brassicaceae* family, Asian greens grow best in cool conditions (the heat gives them a sharper, more bitter taste). However, some varieties are well-suited to summer growing. Mustard greens put up beautiful edible yellow flowers, which we gather into tasty bouquets that we sell at market. With the exception of Chinese cabbage, which can keep many months in the cold room, Asian greens are not a good storage crop. They keep best if left in the ground, which gives a certain amount of latitude in choosing when to pick them.

INTENSIVE SPACING: 3 rows 10 inches apart, spaced every 12 inches

PREFERRED CULTIVARS: Monument (Chinese cabbage), Black Summer (bok choy), Hon Tsai Tai (mustard greens), Tatsoy (as baby heads)

FERTILIZATION: Light feeder

DAYS IN THE GARDEN: 40 to 60 days after transplanting

NUMBER OF SEEDINGS: 4 (April 1, April 15, July 8, August 25)

Kale and Swiss Chard (Brassicaceae and Chenopodiaceae)

Although kale and Swiss chard belong to different families, we grow them in essentially the same way. They are the two main leafy greens that we grow throughout the season, with little effort and very few troubles. Of the two, kale is by far the more popular crop due to its delicious taste and high nutritional value. For some people, eating kale is almost political. Remarkably, its popularity has risen tenfold in the last few years.

Both Swiss chard and kale are semi-perennial crops, meaning that leaves from a single plant can be harvested several times a season. We stagger the Swiss chard and kale seedings so that we can alternate them in our shares and have some of each to sell at market each week. Spring kale is followed by Swiss chard in the summer and a subsequent crop of kale in the fall. This succession takes advantage of the qualities of each plant. Kale grows well in cool conditions, which makes it a good spring crop, but its cold-hardiness is even more impressive at the end of our season: it can survive temperatures as low as 14°F (−10°C) and can stay outside until the last harvest of the year. Swiss chard aptly fills the gap between these two crops, as it does not go to seed in the heat of the summer and can be harvested all season long.

Kale and Swiss chard are both light feeders and do not usually require much upkeep. Even so, flea beetles can cause major damage on young kale plants and even mature ones during hot, dry summers. We use a floating row cover systematically for this reason. When it comes to diseases, Swiss chard requires more attentive care as it can suffer from Cercospora leaf spot, a fungal disease also found in beets. If we detect the fungus early enough, we remove all affected leaves, which limits the progression of the disease. If the infestation is widespread, however, we intervene periodically with a fungicide until the last harvest.

We harvest both of these leafy greens by simply twisting the outer leaves away from the base of the plant and bunching them together. By picking regularly, you can ensure that the plant will always form new tasty leaves, and will keep producing for a very long cropping period. Generally speaking, leaf vegetables must be cooled soon after being harvested, and will keep for over a week in the cold room.

INTENSIVE SPACING: 3 rows 10 inches apart, spaced every 12 inches
PREFERRED CULTIVARS: Red Russian (spring kale), Dinosaur (fall kale), Bright Lights (Swiss chard)
FERTILIZATION: Light feeders
DAYS IN THE GARDEN: ± 90 days, including 5–6 weeks of harvest
NUMBER OF SEEDINGS: 3 (April 9, June 10, July 5)

Kohlrabi (Brassicaceae)

Kohlrabi is an odd-shaped vegetable that makes for a great conversation piece at our market stand. It definitively looks weird, but it's really tasty and nutritious. It took a while to educate our clients about it, but every time we would give out free kohlrabi samples, the reaction was pretty much the same—wow, this thing tastes good!

From a growing perspective, kohlrabi is also pretty interesting. It's a fast-growing, cool-season crop that is ready to harvest just a few weeks after transplanting. We plant only one crop in the spring and another in the fall, aiming to have it ready for the first and last CSA shares and markets of the year.

Kohlrabi can be planted very densely, is easy to weed, and not difficult to grow if the plant isn't stressed. A hot and dry season can affect the bulb, causing it to become woody or spicy like a radish. We keep our first transplant under a row cover until daytime temperatures reach about 75°F (23°C). The row cover also protects the crop from flea beetles, which can be a threat in the spring. Because it is susceptible to attack by the Swede midge, we grow our second seeding under anti-insect netting. We plan the kohlrabi bed to be located beside a broccoli or cauliflower bed so the same piece of netting can be shared between the two.

We harvest kohlrabi when the bulb is two to three inches in diameter and sell it with its leaves (which are edible) in bunches of three. For longer storage (it will easily keep for a few months), we remove the tops and place it in the cold room in sealed bins.

INTENSIVE SPACING: 3 rows 10 inches apart, spaced every 8 inches

PREFERRED CULTIVARS: Korridor (standard), Kolibri (purple), Kossak (fall, large and better after a frost)

FERTILIZATION: Light feeder
DAYS IN THE GARDEN: ± 40 days after transplanting
NUMBER OF SEEDINGS: 2 (April 10, July 1)

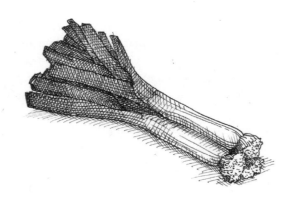

Leek (Liliaceae)

It took us a few seasons to figure out how to grow great leeks. Not that it's difficult or problematic, it's just that most of our customers only eat the white part of the stem. After having that realization, we took on the challenge of producing leeks with nice long white shafts.

We start our leeks very early in the spring. Along with tomatoes, they are the first seedlings started in the greenhouse. We do this not to harvest them earlier (leeks don't really sell before fall is looming) but because we want our transplant seedlings to grow long. We start them much the same way as we do onions, with two exceptions: we do not trim the leaves while they are in the flat, and we use deeper containers to allow for increased root development. We do only one leek seeding (but use varieties with different numbers of days to maturity) and transplant the seedlings into the gardens in early May. This gives them about 10 weeks to grow to a pencil-size thickness and at least 10 inches long, which is what we want.

The more of the leek that is underground, the more white you will get. Usually this is done by gradually hilling soil up against the growing leek as it gets bigger. This way of working requires extra soil on either side of the rows, thus allowing only two rows in a 30-inch bed. We do things differently. Instead of hilling our leeks, we bury them at the transplanting stage, which allows for three rows per bed without any hilling at all.

When our seedbeds are prepared (broadforked and firmed with the roller of our rotary harrow), we use a gouge to make holes 1 inch wide and 8 inches deep. To make this task more efficient, we asked an artisan friend to make us a broadfork-style punching dibbler that makes 3 holes at the same time, while marking the next ones at six-inch intervals. We then take individual transplants, cut their roots to about one inch and simply drop them in the hole, leaving everything just as it is. We don't fill in the hole—and this is very important. Instead we water the leeks with a garden hose so that a little earth is washed down onto the roots. Over time, the first few hoeings will fill the pockets as the transplants get bigger. The first years we tried this, we were worried that having only two inches of the plant poking up wouldn't be enough for it to get adequate light, but it is. After a week, if we see that some transplants did not make it, we use a single dibble (a hand drill with a one-inch mesh works well) to fill in the rows with leftover plants.

The result of all of this work is a guaranteed 8-inch elongated blanched stem. The leeks are not the biggest they can possibly be, but going from two rows to three is an absolute gain in yields per square foot. With our fall leeks, we usually enhance this effect even more by mulching the leeks with straw halfway in the season, which blanches them even further up the shank. Grown in this way, we manage to produce a leek with almost 1 foot of snowy white shank. Customers love this, and we get to be the talk of the town every time.

Leeks are less susceptible to disease than onions, and in our garden we have faced no enemies other than the leek moth. The larvae of the moth dig tunnels through the shafts, making the vegetables unsellable. Since we know this insect is always present in our gardens, we cover our leeks with anti-insect netting when the egg-laying season is in full swing, i.e., late August and early September. Besides using the nets, the only other maintenance required is regular hoeing, since keeping the leek beds weed-free is essential with such intensive spacing.

Leeks are one of the most flexible crops you can grow. They can be harvested at any stage of development and are frost-hardy enough to stay in the ground until the end of the season. For both summer and autumn leeks, we harvest them when we feel they are big enough (about 1½" in diameter) and when we have enough demand. We tie them up and sell them in bunches; the number per bunch depends on the size of the vegetables. For the purpose of presentation, we trim the roots and cut off the tips of the upper leaves in the shape of a chevron. Leeks keep well in the cold room for a few months.

INTENSIVE SPACING: 3 rows 10 inches apart, spaced every 6 inches

PREFERRED CULTIVARS: Varna & King Richard (summer), Megaton (fall)

FERTILIZATION: Heavy feeder

DAYS IN THE FIELD: 75 to 130 days after transplanting

NUMBER OF SEEDINGS: 1 transplanted in early May

Lettuce *(Asteraceae)*

Head lettuce is the most popular crop we grow. Almost all our customers buy some at the market, and our CSA families appreciate having one or two every week in their share. In terms of profitability, lettuce is right up there with cucumbers and tomatoes. For the market gardener, lettuce is a high-yield, high-reward crop that is not complicated to grow. The biggest challenge with lettuce is producing it every week without having different plantings maturing at the same time.

We've developed a planting schedule where we start new seedlings every 15 days, each time choosing two cultivars with different growing periods (e.g., 45 and 52 days respectively). This takes the guesswork out of succession planting and makes certain that we will have staggered lettuce crops. When doing our starts, especially during the summer, we also plant 30% more seedlings than necessary, a safety factor that offsets any unforeseen germination or plant failures at the transplanting stage. Lettuce seeds are generally inexpensive, so this insurance approach is worth it. With regard to maintenance, lettuce follows straightforward procedures. We plan for a false seedbed before planting, and use our collinear hoes (which nicely get in and around the lettuce heads) to keep the crop weed-free.

Lettuce sometimes falls prey to slugs and tarnished plant bugs, but never severely enough to call for an intervention. In some years, we have had mildew spontaneously appear in the garden and wipe out several plantings. This has not been an ongoing problem, but we are still looking for the best solution if this begins to happen more regularly.

One important thing to know is that lettuce is highly sensitive to lack of water, going bitter in overly dry conditions. Our customers are not too pleased when this happens and are not afraid to speak up about it. As a result, lettuce has become something of a "barometer" crop that tells us when to irrigate our gardens. We keep rain gauges in the lettuce beds. If we see that there has not been enough rainfall in a given week, we do not hesitate to turn the sprinklers on. Regardless of being well irrigated, lettuce can also bolt when temperatures get too hot. We avoid this by choosing the best-adapted cultivars and implementing these choices in our planting program. We also do trials every year to see which ones we like best.

On harvest days, we always start with lettuce, as it is more vulnerable to the heat of the day than other crops. We cut the heads off at the base with a knife, dunk them in a cold water bath, and let them drip-dry. We store the lettuce in a closed bin in the cold room, where it can stay fresh and crisp for about a week.

INTENSIVE SPACING: 3 rows 10 inches apart, spaced every 12 inches

PREFERRED CULTIVARS: Salad Bowl (red or green oakleaf), Jericho (summer romaine), Nevada (Batavia), Buttercrunch (Bibb), Vulcan (red), Grand Rapids (curly)

FERTILIZATION: Light feeder

DAYS IN THE GARDEN: 30 to 45 days after transplanting

NUMBER OF SEEDINGS: 8 spaced 15 days apart from mid-April to mid-August

Melon (Cucurbitaceae)

Melons are like the fruit of the gods. Anyone who has tasted a garden grown melon at the peak of its ripeness knows exactly what I am talking about. Its delicious flavor, enchanting aroma, and juicy texture is loved by all our customers, making this low-yielding and space-hogging crop worth the sacrifice. Even if growing melons goes against our "rule" of intensifying production wherever we can, we grow them with enthusiasm. We really have no choice—our customers would never forgive us if we didn't!

We seed melons the same way we do summer squash. In fact, the two crops are similar in lots of ways, and the reader may refer to the crop notes on summer squash for tips on planting and plant protection. Despite these similarities, melons are more sensitive to temperature than summer squash, and like cucumbers, they do best when transplanted into soil that is very warm. This is why we seed our melon crop relatively late in the season and do so in plastic mulch. Since melon plants are quick to sprawl out, we use a large piece of landscape fabric that covers both the bed and the pathway, which keeps weeding to a bare minimum.

Knowing when to harvest the fruit is really important to ensuring optimal enjoyment. If a melon is picked too green, the flavor will be highly disappointing; if picked too ripe, the fruit is too watery. To determine the perfect moment for picking, you need to look for certain signs on the fruit. Honeydew melons and cantaloupes are ready as soon as they start to turn yellow. The peduncle forms a ring-like crack or becomes chapped, indicating that the fruit is ready to break off. This is when the fruit starts to smell like a melon, and you can really sense that it is ready. Melons that break off on their own often turn out overripe. (This is not the case with watermelons however: ripe watermelons make a hollow sound when tapped.) If you turn a melon over and notice a pale patch where the fruit was sitting on the soil, this is a good sign. The patch turns from pale to yellow as the fruit matures. But tapping and turning aside, the best way to tell if a melon is truly ripe enough is to taste it!

Melons harvested at the right time keep at room temperature for a week at most. If we have melons that were harvested overripe, we keep them in the cold room to avoid spoilage—although this saps them of much of their aroma and flavor. Watermelons stay fresh for about two weeks at the temperature of our storage room.

INTENSIVE SPACING: 1 row, spaced every 18 inches
PREFERRED CULTIVARS: Cavaillons (Charentais), Sivan (Charentais), Halona (muskmelon), Sweet Beauty (watermelon)
FERTILIZATION: Heavy feeder
DAYS IN THE GARDEN: 65 to 85 days after transplanting
NUMBER OF SEEDINGS: 1 (May 24)

Mesclun Mix (Various Families)

The word "mesclun" comes from *mesclar*, a word in the Occitan language spoken in Provence. This word is derived from the Latin *misculare*, meaning "mix thoroughly." In France, especially in the south, mesclun is defined specifically as a mix of baby lettuce, chicory, arugula, and sorrel. In Quebec, however, we have no rigid ideas about what mesclun should be, and ours includes any number of leafy vegetables of various colors and textures that are pleasing to the eye and palate. The only criteria are that they should be bite-sized (two to four inches long) and prewashed, ready to be eaten. Our "recipe" calls for Asian greens in the spring and fall, with the first and last seedings grown in unheated hoophouses, and a mix of lettuces in the summer. Throughout the season, we also add baby Swiss chard, chicory, or kale to supplement the basic ingredients.

Mesclun is a crop designed for the market garden: it's fast growing, so that in just 30 to 40 days it generates a good revenue per square foot; it's a great catch crop, allowing rapid turnover for succession seeding; and you can grow it pretty much year round, even in our harsh winters. For all these reasons, we've made mesclun our signature crop and put a lot of effort into growing the very best. At present, it's the only crop that we sell wholesale; and in doing so, we guarantee a steady supply every week to our local grocery store, as well as to several restaurants. Here lies the biggest challenge with this crop: producing beautiful meslcun in sufficient quantity to meet weekly sales, and doing so without any shortages due to crop failure, or variations in growth due to changing weather. We manage this task by planning for a new crop every 15 days, aiming to have two or three cuts from the same planting. This semi-perennial approach enables us to buffer production and to pick and choose what we want from different seedbeds on a weekly basis. Our seeding dates are closer together in the fall, as we need to take into account the slower growth of the greens as the days get shorter. Light, not temperature, is the limiting factor when extending the mesclun growing season.

We direct-seed our mesclun mix using a Six-Row Seeder. This tool seeds a 12-row bed in two passes, which gives us a very intensive crop (about 45 pounds per 100-foot bed, depending on the greens in the mix). While this technique gives us excellent yield, it does have its drawbacks. Firstly, planting 100 feet of salad consumes up to 2.5 ounces of seeds, which means we rack up some pretty pricey seed orders. Next, the beds must be kept weed-free, as hoeing is impossible with such dense seedings. To keep weeding to a minimum, we prepare our beds for seeding well in advance and always use the false seedbed technique. When the leaves reach a height of about 1 inch, we use a hand-pushed finger weeder that cultivates the exposed soil between the rows. This provides the crop a good head start over the weeds, eventually forming a leafy canopy that shades out undesirable weeds.

When it comes to diseases and insect pests, our biggest concern is the flea beetle, which can ruin a crop if not kept under control. Breaking up our plantings of Asian greens with lettuce is generally enough to interrupt the pest's life cycle, but we still cover the mesclun beds with row covers or insect nets immediately after seeding as an added protection. In addition to this preventive measure, we do regular inspections. In the past, we have actually trapped flea beetles inside the netting and been forced to use insecticide. Vigilance is key.

We used to harvest our greens with field knifes, cutting one handful at a time. Harvesting about 300 pounds a week this way is feasible, but it's time-consuming and takes a toll on your knees and back. We now use a hand-held electric salad mix harvester powered by a hand drill which swiftly cuts the greens at their base. This simple technology helps us get the job done neatly and quickly, cutting the work by more than 80% (a weekly harvest used to take three people working for three hours; now it takes one person less than two hours). Needless to say, this harvester was a great investment.

In addition to this gain in productivity, we've also learned to better select the plant varieties that weigh more, and that are tougher in face of washing and handling. The main ingredients in our mesclun mixes are baby leaf romaine, oakleaf lettuces, kale, Swiss chard, and chicory. We also harvest from a couple of beds of mini heads of lettuce, from which we harvest three to four time, as well as other uncommon garden forage: Brassicaceae flowers, pea flowers, overripe lettuce cores, and young cabbage sprouts. Our clients really appreciate the creative diversity but not as much as the quality of each ingredient. We don't pick greens that have become too thick, too spicy, too coarse, or simply too ugly. When it comes to competing at the market, our high standard is hard to beat.

Once the leaves are harvested, we put them in a cold water bath where we gently mix them. We take this opportunity to remove any damaged leaves, insects, and weeds. Then we spin the greens in in a washing machine, a very important step if the salad is to have a long shelf life. We then put the mesclun in half-pound bags identified with our farm's logo. Although it comes with a higher price tag, 9 out of 10 times people chose it over California salad mix, which doesn't last as long. Our freshly harvested artisanal mesclun mix will stay fresh for over a week.

INTENSIVE SPACING: 12 rows 2¼ inches apart, spaced every ½ inch

PREFERRED CULTIVARS: Asian greens (Ruby Streaks, Tatsoi, Mizuna), lettuce (Tango, Buttercrunch, Lollo Rossa, Firecracker), arugula, kale (Red Russian), Swiss chard (Rainbow), spinach (Space), Salanova (Mini-head)

FERTILIZATION: Light feeder

DAYS IN THE GARDEN: ± 45 days for two cuttings
NUMBER OF SEEDINGS: Outdoors—every 15 days from mid-April to mid-September. In tunnels—March 5, March 10, March 20, March 28, September 25, October 5, October 10.

Onion (Liliaceae)

Onions are a crop that we want to grow plenty of. They are widely consumed by most families and can be sold throughout the whole market season. As a staple food, however, onions often sell for marginal prices. We compete with bigger commercial growers by offering interesting diversity. Popular onions such as shallots and cipollinis aren't found in most grocery stores, but food lovers in our region know they can find them at our market stand. To meet the high demand, we begin with scallions and green salad onions (two of the highest-yielding crops) and eventually move on to fresh summer onions. Toward the end of the season, many of our customers will stock up on cured storage onions for the winter.

We start all of these onions at the same time, at the end of March. We sow the seeds by broadcasting onto open flats, which saves us space. As the onions grow, we clip their leaves once or twice to a length of 4 to 5 inches, to stimulate their development. We transplant them in early May into soil fertilized with nitrogen-rich chicken manure. When transplanting, it is important to ensure that both the soil mix and the ground are wet enough, as moisture helps the tiny onion plants get established. They should be planted as close to the surface as possible, as they do not like being put too deep in the soil.

We used to set out young onions at a spacing of 2 to 3 inches apart in each row, like most growers do. Now we now transplant them at wider spacing, but in clumps of 3 or 4 instead of one at a time. Besides providing more yield, this space-saving technique makes the long job of transplanting onions go much faster. Planting onions three to a hole also makes the crop much easier to hoe later on. While this spacing may seem tight, the onions will have no trouble developing to a robust size, given that the soil is loose and hoed regularly. For early harvests, we cover many of our onion beds with row cover (supported by hoops) immediately after transplanting. This protection makes an enormous difference in the early stages of growth and really makes the vegetables take off. With onions, the secret to success is establishing the plants well and encouraging vigorous leaf growth first and foremost.

The main challenge with this crop is weed control. For one thing, onion leaves don't shade out weeds, and once the onions are more than 8 inches tall, they become hard to hoe without damaging the plants. Moreover, the presence of damp weeds, especially Galinsoga, increases the risk of fungal and bacterial diseases. By frequently hoeing and hand-weeding in the early stages of growth, we manage to keep our onions relatively weed-free. It's worth mentioning that the better you manage weeds on the plots that precede the onions (the year before), the easier it will be to keep weeds in check the following year. This is something to keep in mind when designing a crop rotation.

The onion maggot is the only real pest that challenges us. But since we keep most of our onions under row covers during the maggot's egg-laying period in May and early June, we have never experienced serious damage. Onions are also vulnerable to many fungal diseases including downy mildew and botrytis, which claimed much of our onion crops in our first few seasons. Since then we have learned to apply weekly applications of copper and sulfur in alternation at the first sign of fungal disease, especially during excessive wet weather or when the leaves have been damaged by hail. In recent years we have also experimented with a biofungicide that inoculates the soil and the plants with a bacterium (*Bacillus subtilis*) to thwart the development of pathogenic fungi.

Summer onions can be harvested at any stage of maturity, which allows some leeway about when to harvest them. If they don't sell on a given week, we can always cure and sell them at a later date. Storage onions are more challenging since they need to be harvested and cured at the right time. These onions are usually ready when their leaves start to yellow and fall down. We pull them from the soil and cut their stems, leaving slightly more than 1 inch above the neck. Then we lay them out on the ground and let them cure (dry) in the sun for a few days before putting them in storage, where they continue to dry. (This is basically the same process that we use for garlic.) The onions are completely cured when their necks are no longer green and are totally closed up. We pack them in mesh bags, making sure to include a mix of different sizes and to remove any that are damaged. Bruises will cause disease, so be careful when handling them. Well-cured onions, handled with care, will keep for 4 to 7 months in a cool, dark place.

INTENSIVE SPACING: 3 rows 10 inches apart, spaced every 10 inches

PREFERRED CULTIVARS: Purplette (early, green salad onion), Sierra Blanca (sweet and large, sold fresh), Ailsa Craig (Spanish onion), Redwing (red, storage), Gold Coin (cippolini), Ambition (French shallot)

FERTILIZATION: Heavy feeder

DAYS IN THE GARDEN: 50 to 110 days after transplanting, depending on the cultivar

NUMBER OF SEEDINGS: 1, transplanted in early May

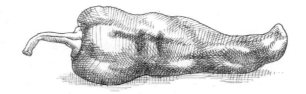

Pepper (Solanaceae)

People love peppers, especially red ones, which children eat with the same delight as they would an apple or a melon. To meet our needs, we grow many different ones, aiming to have them ready as early as possible. When red bell peppers can be harvested by the end of July, this high-yielding crop becomes a real money-maker. To get top dollar for our peppers, we choose cultivars that produce medium-sized fruit. We know from experience that most people will scowl at paying $3 for one large pepper, but are quite fine with paying $4 for two smaller ones.

We grow most of our bell peppers inside hoophouses to ensure an early harvest. We start our seedlings the same way as our tomatoes, bringing them on to about 8 weeks old before putting them in the ground. They are transplanted into warmed-up soil (see cucumbers) and set to be watered by a drip line that is irrigating daily to the equivalent of an inch or so of rain per week. Peppers are lighter feeders than tomatoes and cucumbers. One application of a mixture of vermicompost and chicken manure is enough to meet their nutritional needs.

In the hoophouses, we use 5' stakes every 2 or 3 plants, with string weaving between them to support the plants when they are loaded with fruit. We also prune the suckers multiple times in the early stages, removing the first flower/fruit to allow the plants to establish a better root system before they start producing in earnest. By mid-August, the plants' heads are removed (i.e., the upper part of each main stem is cut above the last fruit) so that the production of new fruit will cease. This allows the plant to put its energy into ripening the already established peppers.

One of the common problems with peppers is blossom-end rot, which appears as a black or beige spot at the bottom of the fruit and makes them unsellable. Blossom-end rot is not a disease as such, but rather a physiological disorder resulting from calcium deficiency, which in turn is caused by water stress during periods of rapid growth. To counter this situation, we add weekly calcium supplements to the drip irrigation water during the fruiting months when the peppers are growing rapidly (July to mid-August). This technique, known as fertigation, is quite simple. It just requires an injector that gradually releases doses of soluble calcium to the plant. The apparatus connects to standard outdoor plumbing.

As far as insect pests are concerned, peppers, like eggplants, are susceptible to the tarnished plant bug. However, ever since we have sealed our hoophouses with insect netting, these critters have stopped being a problem. The growing conditions of the hoophouse, however, may bring about an increase of aphids, which can be problematic. We then scout our plants on a weekly basis, and if we find that the aphid population is developing too strongly, we introduce ladybug predators (which can be purchased online and delivered by post). While waiting for them to be delivered, we spray pyrethrum on any plant that may be harboring aphids.

Given that ripe peppers do not keep for long in the cold room (only about 10 days), it's best to sell them as quickly as you can. Once pepper production is underway, we harvest the nicest and reddest fruit twice a week. In addition to sweet pepper plants, we grow a few hot pepper plants to please the spice aficionados (especially ourselves!).

INTENSIVE SPACING: 1 row, spaced every 9 inches
PREFERRED CULTIVARS: Orion (very early, not too big), Carmen (Italian-style, delicious), Round of Hungary (ribbed), Mandarin (yellow), Hungarian Hot Wax (hot)
FERTILIZATION: Heavy feeder
DAYS IN THE GARDEN: Entire season
NUMBER OF SEEDINGS: 1, transplanted end of July inside a hoophouse

Radish (*Brassicaceae*)

Believe it or not, radishes have become one of our top sellers at the market. This may have something to do with the cultivars we choose: customers are often dazzled by piles of vivid, brightly colored radishes that stand out at our booth. Radishes are easy to grow, quick to harvest, and are the only vegetable that we intercrop. Early staggered plantings of radishes nicely fit at the base of slower crops such as summer squash, cucumber, and pea.

Like most other crops in the Brassicaceae family, radishes thrive in cool conditions, so we avoid growing them in mid-summer. The heat of summer causes the radishes to turn spicy and go to seed early, which in turn makes the root go stringy. We therefore plan our seeding for spring, late summer, and fall harvests. We also plan for winter radish, a delicious, colorful, and frost-tolerant crop that few people know about. When grown under row cover, these radishes can stay in the garden very late into the fall, right until the last of our vegetable shares.

The only concern with growing radishes is that, if left unprotected, they almost always get ravaged by the cabbage maggot and its larvae, which make black marks in the root. Flea beetles can also be a problem in a hot, dry spring. We therefore always cover our radish crops with anti-insect netting or row cover immediately after planting.

We pick our radishes once they have reached a medium size, about two inches in diameter. Although the harvest can be spread out over two weeks, it is better to harvest them small so they do not become spongy, split, or coarse. We sell only the best-looking radishes in bunches of 6 to 12. We also mix colors of different cultivars to increase their appeal.

INTENSIVE SPACING: 5 rows 6 inches apart, spaced every 1¼

PREFERRED CULTIVARS: Raxe (spring), Pink Beauty (pink), French Breakfast (long, European), Red Meat (winter)

FERTILIZATION: Light feeder

DAYS IN THE FIELD: ± 45 days, including 2 harvests

NUMBER OF SEEDINGS: 4 (May 10, May 23, July 11, August 20)

Snow/Snap Peas (Fabaceae)

Snow and snap peas are eaten pod and all—often raw for maximum taste and nutritional value. Their extraordinary garden-fresh flavor is always a hit at farmers' markets, where their appearance signals the beginning of a new vegetable season. Like beans, peas require considerable work, not only during harvest but also while setting up stakes and trellising them as they grow. Therefore, we plan our pea and bean plantings to avoid having to pick both at the same time. We also don't shy away from selling peas at a premium. If ever we get gripes from customers, we simply remind them that our peas are not imported from China and that the price is only what we need to ask in return for growing them.

Because peas germinate best in cool soil, we do our first pea seeding quite early in April. We seed them in just one row, by hand, spaced very tightly together, which we then cover with row cover. Seeding only one row per bed is not optimal in terms of yield per square foot, but it makes both harvesting and weeding much easier.

We prefer indeterminate cultivars to bush varieties because they last longer and produce more (and tastier) peas, even if trellising them requires additional effort. To encourage the plants to climb, we support them by tying a cord to stakes spaced every 15 feet along the length of the bed. We tie a new cord higher every week so that the crop will stay contained as it grows.

Snow and snap peas are easy to grow and have no real pests. To limit the development of fungal diseases, it is best to avoid harvesting them when the plant leaves are wet with dew or rain. For maximum taste, we pick them when the peas fill the pod and the pods are nice and round. If you wait too long, the peas will become stringy and tough. The crop must be harvested every two or three days for best results. To harvest as efficiently and ergonomically as possible, it is imperative that you pick peas with both hands and stay focused; this is a message we are always hammering into our harvest helpers. Snow and snap peas remain crunchy for about a week in the cold room and therefore must be sold quickly.

INTENSIVE SPACING: 1 row, spaced every ½ inch along the row
PREFERRED CULTIVAR: Super Sugar Snap (fabulous)
DAYS IN THE GARDEN: ± 85 days, including two or three weeks of harvest
FERTILIZATION: None
NUMBER OF SEEDINGS: 2 (April 19, May 13)

Spinach (Chenopodiaceae)

Any market gardener growing crops in Quebec-like winters will appreciate the value of spinach. Not only is it the ultimate frost-hardy crop (surviving temperatures as low as 19°F/−7°C), but unlike Asian greens and kale greens, it is known and loved by all.

Our seeding dates are set so that we get a few harvests of spinach before and after the summer season's lettuce crops. While the demand for this vegetable is high all through the year, it is difficult to grow in the middle of summer: the plant tends to go to seed early when the days are long and the weather is hot (starting in July). Besides, the plant requires cold temperatures to grow the sweet, tender spinach we love best.

Although spinach is usually direct-seeded, its germination rate is highly variable, so we prefer to transplant it. This means extra work, but it is worthwhile: the perfect spacing obtained from transplanting ensures optimal yield. We always cover our first spring seeding with row cover to make it develop as early as possible.

We sell spinach in bulk. At harvest time, we pick the biggest leaves of each plant rather than harvesting the whole bunch. This is finicky, but it also gives us the highest total yield per plant. When washing spinach, it is important to spin-dry it well, as it must be completely dry in order to stay fresh in a sealed bag. In these ideal conditions, bulk spinach keeps for over a week in the cold room.

When we want to include spinach in our mesclun mix, we grow it differently. Spinach for mesclun mix is planted with the Six-Row Seeder by filling every other hopper (i.e., six rows per bed). The reader may refer to the crop notes on mesclun mix for the harvest and washing process, as these are the same for spinach. For late crops, the seeding dates must take into account the diminishing day length in the fall. On our farm, mid-September is the target date for seeding mesclun spinach in a cold tunnel if we plan to do one more harvest.

For some customers, bulk spinach immediately calls to mind *E. coli* contamination. Simply remind concerned customers that spinach is no more susceptible to infection by these fatal bacteria than any other raw vegetable.

INTENSIVE SPACING: 4 rows 6 inches apart, spaced every 6 inches along the row
PREFERRED CULTIVARS: Space (smooth leaf, spring crop), Tyee (curly leaf, fall and spring crop)
FERTILIZATION: Light feeder
DAYS IN THE GARDEN: 30 to 50 days after transplanting, including 2 or 3 cuttings
NUMBER OF SEEDINGS: 4 (April 1, April 25, July 25, August 5)

Summer Squash *(Cucurbitaceae)*

Most people know this long, dark green vegetable as zucchini. It is actually the same vegetable, but summer squash has more variety in shape and color, and is harvested smaller. Summer squash are easy to grow, profitable, and highly productive. You often hear stories about CSA partners not knowing what to do with all the abundance, and so not planting more than you need is definitely good advice.

We plan for 3 seedings in our crop plan. Our first one is done very early inside a hoophouse, so as to have our first fruit as early as the end of May or beginning of June. Our second seeding is done a few weeks later directly into the garden, and is the trickiest to grow. Summer squash is a heat-loving plant that is not very tolerant of the cold, and even less tolerant of frost. To help them along, we cover the crop with floating row cover hung over hoops. However, the hoops must be installed high enough to allow for the size of a mature summer squash plant, which increases the greenhouse effect inside the mini-tunnel. This does a great job of heating up the air and the soil, but it can also cause heat shock to transplants that are still fragile. To prevent this, we dip our plants in a kaolin solution when transplanting. This soluble clay powder, a natural and biodegradable product, considerably reduces transpiration in young plants and helps them acclimate to the heat. We also make sure to irrigate and monitor the crop during sunny weather, sometimes uncovering the crop to properly ventilate it.

The striped cucumber beetle finds young summer squash plants particularly tasty and can cause major crop damage. The row covers do a good job of protecting the plants, but at the flowering stage, the plants must be uncovered to allow pollinators to access the flowers, thereby ensuring proper fruit formation. At that time, the plants are usually sufficiently well-developed to not fall prey to cucumber beetle predation; however, the beetles also spread bacterial wilt, a disease that causes the plants to wilt within a few days of being infected.

We then manage this insect by burning them using a propane blowtorch while they are inside the flowers of the plant. We do this only on the male flowers (which won't affect the production yields) and do so very early in the morning, while the insects are still relatively inactive and the bees have not begun their foraging at the squash flowers. Repeating this strategy several times per week greatly reduces the population pressure, but really all it does is delay the inevitable. After five to six weeks of good harvest, most plants will succumb to bacterial wilt. This is the main reason why we plan another succession in the middle of the summer. In any case, this relay planting is a good practice since summer squash yields are best in the first 2 months of production.

We harvest summer squash when they are six to eight inches long, depending on the cultivar. They are the most tender at this size and command a better price at the market. Summer squash must be picked every two to three days; this prevents the fruit from becoming bulbous and stimulates the growth of new ones. To harvest the crop as efficiently as possible, we wear a tree-planting bag. The bag lets us carry the harvested squash and keep our hands free as we walk the pathways. Squash flowers are highly coveted in fine restaurants, and our market customers often come looking for them. We set aside a few summer squash plants to be devoted exclusively to flower production. The flowers must be picked at dawn while the flower petals are starting to unfurl in the sun. This adds to our workload in the morning, but it is worthwhile because of the price some people are willing to pay.

Summer squash stays firm in the cold room

for about a week. The flower is very delicate and must be sold on the day it is picked.

INTENSIVE SPACING: 1 row, spaced every 24 inches
PREFERRED CULTIVARS: Plato, Zephyr (very beautiful), Sunburst (pattypan), Portofino (delicious), Costata Romanesco (prolific production of male flowers)
FERTILIZATION: Heavy feeder
DAYS IN THE GARDEN: 70 days after transplanting
NUMBER OF SEEDINGS: 3 (April 4, May 3, June 20)

Tomato (Solanaceae)

"Are these *real* tomatoes?" We get asked this question at the market on a regular basis by newcomers who are not familiar with the reputation of our farm. Grocery store tomatoes might look good, but they are tasteless by comparison. Come summertime, more and more people are on the lookout for local flavorful tomatoes, so growing them is guaranteed to attract customers to the market stand. The advantage that the small market gardener has over large-scale producers who sell through the supermarket chain is that we have the ability to harvest tomatoes that have fully ripened on the vine. Our tomatoes don't keep as long as the other guys, but they sure do pack a lot of punch in the flavor department. Since they are so popular (especially early in the season) and given that most people are happy to pay extra for garden tasting tomatoes, they are the most profitable crop you can grow.

We have not grown field tomatoes in a number of years and dedicate the entire production of our heated greenhouse solely to this crop. This approach eliminates most diseases commonly found with outdoor operations and extends our growing season by almost two months, allowing us to harvest as early as mid-June and as late as mid-October. Growing tomatoes in a controlled environment encourages an increase in both the quantity and quality the fruit, but above all, by growing indeterminate tomato plants using greenhouse growing techniques, tomato production yields can be increased tenfold. Although this practice requires a certain capital investment and is labor-intensive, the economic payback is well worth the investment. It should be noted that when I refer to greenhouse tomatoes, I mean tomatoes grown in living soil, not those raised hydroponically.

Production of greenhouse tomatoes is not without serious challenges, however. So many technical factors are involved in the practice that it should be considered a trade unto itself. On our farm, one person is responsible for directing the operations, and we've even hired an agronomical advisory service specializing in greenhouse production to help us deal with the complex parameters involved. This being said, while the techniques we use in our greenhouse are sophisticated, the greenhouse itself is not. This is an important point, as some greenhouse growers pay a fortune for all kinds of bells and whistles to control and improve production. By contrast, we ventilate our greenhouse with roll-up sidewalls that we manually manage, and the only two programmable tools we use are a thermostat and a timer to automatically turn on the drip irrigation. Growing great greenhouse tomatoes in a modest facility is feasible if you know what you are doing.

Like I mentioned, we plan to have tomatoes by the middle of June, when their value is at their peak. To accomplish our goal of an early harvest, we begin our seeding at the beginning of February and transplant the seedlings by the first week of April. One of the keys to a successful tomato crop is high-quality transplants, and we do everything we can to ensure just that. We heat the plant nursery (at least part of it, see Chapter 6) to the optimal temperature (65°F/18°C at night, 77°F/25°C during the day) to encourage optimal growth. The seedlings will be potted up in 6-inch pots to allow maximum root space and as many nutrients as possible. The plants are ready to plant out when they are about 8–10" tall, sturdy, and dark green.

We also graft our tomatoes. This simple yet precise operation consists of cutting two plants with different characteristics (a rootstock with a large root system that is resistant to multiple diseases and a scion which gives the fruit of the desired cultivar) and then splicing them together with a small clip. From then on, the two stems grow together, the cut heals over, and you have a grafted plant giving you the best of both worlds: a sturdy root system and the fruit-bearing plant of your choice. The main reason to graft tomatoes is to avoid soilborne diseases, in particular corky root, which can become a problem when tomatoes are grown every year on the same spot. We've been growing tomatoes in the same greenhouse for almost a decade now and never face any problems. It certainly is simpler (and cheaper) than moving greenhouses around. But even if soilborne diseases were not a problem, grafting plants significantly increases our yield, thus warranting the extra work. Reference materials that detail the exact procedures are easy to find, but the best way to learn how to graft is to learn directly from an experienced greenhouse grower.

In our greenhouse, we've modified the width of our beds to be narrower than the pathways (24-inch beds and 36-inch pathways). This layout makes for easier tomato harvesting and allows us to trellis the vines in a V-shape manner, with the plants starting from the middle of the bed and growing up and out toward the pathway on each side. The overall density is about two plants per square yard, which comes down to spacing plants every 9 inches on the row. This intensive spacing calls for a sturdy trellis system (one plant will hold 10 to 12 pounds of fruits, and we grow more than 100 plants per bed). Our system comprises two steel wires suspended lengthwise above each bed at a height of 8 feet and fastened at both greenhouse end walls to a tightened cable (cable diameter is 1/16 inch). The plants are then trained up a polypropylene twine attached to a hook on the steel wire.

As the plant grows, we keep winding them around the twine, wrapping the plant stem in a clockwise manner. Some growers use clips to do this, but we find it goes faster to work without them. Indeterminate tomatoes require pruning of all lateral branches (suckers) as they develop in order to encourage a single leader. We do this every week so that only small side shoots are removed. We also plan this operation for sunny days, so that the wounds heal easily, thereby decreasing the likelihood of the plant developing infectious diseases. Every two weeks, we trim the tomato clusters to maximize the size of each fruit and optimize the balance of the plant. At that time, we also remove old leaves from the lower part of the plant to improve air movement under the plants and to make the growing plants easier to manage. To ensure proper pollination of the tomato plants, we walk along the plants every couple of days, hitting the steel wires with a stick every 5 feet or so.

As the plants reach the trellis cable, we lower and lean the plant about one foot further out to the side so that it can be maintained within the trellis system. Eight weeks before the last harvest, we cut off the heads of the plants to allow the existing fruit to reach full maturity. These maintenance practices are standard greenhouse procedures, and more detailed information for every step of the way is not hard to find.

Tomatoes produce the largest yield of highest-quality fruit when daytime temperatures are in the range of 80° to 86°F (26° to 30°C) and when night temperatures remain above 65°F (18°C). As I mentioned above, we don't have computer controllers to maintain these temperatures at all times, but we do manage to keep the temperature in this range by setting up an alarm thermometer that informs us when temperatures are getting off target. Temperatures need to be lowered on cloudy days, when plants don't have enough sunlight to allow sturdy growth.

Sufficient fertilization input is of primary importance to meet the needs of the plants over time. Each month, we fertilize with a mixture of vermicompost, poultry manure, and potassium sulfate that we integrate into the soil with some light hoeing. To ensure that these amendments mineralize properly, it's important that the soil be kept moist. For this reason, we cover the ground with 4-foot strips of UV-treated silage tarps that cover both the pathway and half of the beds (right to the base of the plants). Each side of the polyethylene is a different color. The white side faces up to reflect sunlight back up to the plants, and the black side faces down towards the ground. The earthworms love the enhanced darkness, and they play a large role in loosening the soil structure for the tomato plants. We are reminded of their valuable contributions each time we remove the tarps for our monthly compost topdressing.

Tomatoes don't usually suffer terribly from pests. In early spring, we may face an aphid problem, but soap spray quickly remedies the situation. The damage done by hornworms can look frightening, but the problems caused by them are generally minor and not serious enough for us to spray. Diseases, however, can be disastrous when not handled properly. The warm, humid conditions in a greenhouse are great for molds and bacterial diseases. Selecting greenhouse tomato varieties that are resistant to these is essential. Controlling moisture is also very important in order to keep the plants as dry as possible. We manage this by heating the greenhouse every morning for a little under an hour, rain or shine (even in the middle of summer), with the side vents open a crack. This lets out any moisture that has built up overnight and dries off the plant leaves. Since we adopted this strategy, our greenhouse tomatoes remain healthy right through the end of the season.

Once production is in full swing, we harvest tomatoes every two or three days for much of the season. The beefsteak varieties are harvested with the calyx still attached, and vine tomatoes are harvested by clipping the main cluster stem from the plant. Both the calyx and stem of the cluster give off the garden tomato aroma that distinguishes the product. As we harvest, we place the tomatoes in boxes designed for greenhouse tomatoes and stack these on a harvest cart specially sized for going up and down the pathways. When collected in this way, the tomatoes will keep for about one week at ambient temperature. We never refrigerate our tomatoes, as this saps them of much of their texture and flavor.

INTENSIVE SPACING: 1 row, spaced every 9 inches along the row

PREFERRED CULTIVARS: Macarena (big and tasty), Trust (reliable), Red Delight (cocktail), Favorita (cherry)

FERTILIZATION: Heavy feeder

DAYS IN THE GARDEN: Entire season

NUMBER OF SEEDINGS: 2 transplants: one in mid-April and another in mid-May

Turnip (Brassicaceae)

Shhh…turnips are a well-kept secret. The major grocery chains have not yet caught on to selling the sweet varieties of turnip, which customers are always pleased to discover. Since we began farming Hakurei, a Japanese turnip, it has been one of our best-sellers at the market and a favorite in our CSA shares. In France, small turnips known as *rabioles* are highly appreciated, and we have made it a mission to popularize them in Quebec as well. The turnip is a cool-weather vegetable, so it loses its tenderness and becomes spicy in the summertime. For this reason, we seed turnip only in the spring and fall, even though we would be able to sell lots throughout the season. Since this crop is able to survive cold temperatures and even light frost, we often grow it late into the season in one of our hoophouses.

We direct-seed our turnips because they sprout quickly and are easy to weed. However, there are two insects that make life in the turnip patch difficult: flea beetle and cabbage maggot. In the past these invaders have completely wiped out some crops. We have learned not to take chances and always keep our turnips covered with row cover or anti-insect netting. Flea beetles are quite tiny, so it is important to choose netting with a fine enough mesh to keep them out. I recommend a mesh of 0.014 inches. Netting is preferable to row cover as it does not create thermal effects in the summer.

We harvest turnips in a gradual fashion, starting with the largest ones. Our seeder places the seeds fairly close together along the row, so removing the largest ones allows the smaller ones to grow bigger and fill in the spaces. After three weeks of harvesting, the roots become stringy and lose their tenderness. This is why it's usually better to do succession plantings rather than stretching a single planting out over a long period. As with all other root vegetables, the leaves of the turnip stay fresh-looking for up to one week in the cold room. Therefore, turnips should be sold relatively quickly.

INTENSIVE SPACING: 5 rows 6 inches apart, spaced every 1¼ inches

PREFERRED CULTIVARS: Hakurei (eaten raw), Milan (less tender, very beautiful), Scarlet Queen (red)

FERTILIZATION: Light feeder

DAYS IN THE GARDEN: 35 to 50 days, including more than one week of harvest

NUMBER OF SEEDINGS: 5 (April 22, May 6, May 23, August 5, August 25)

The Forgotten Few

As I have mentioned a few times in this handbook, there are a number of major crops that we choose not to grow at all. Rather than describing how others go about producing them, I feel it is helpful to explain why we do not grow them.

Let's start with the potato, a vegetable with all the characteristics of an appealing crop. Potatoes are popular and indeed profitable in terms of surface area. However, the tubers are much easier to harvest with a potato harvester than by hand. For a market gardener, it is almost impossible to be as productive at potatoes as a mechanized grower, and the price of potatoes would not reflect the amount of effort involved. Having said this, new potatoes are something to consider. If you grow them in a tunnel very early in the spring and manage to be the first to sell them at market, these can be lucrative.

Sweet corn is another beloved garden crop. However, it takes up a lot of space in the garden for a small amount of yield. On farms where growing space is not limited, a crop of non-GMO, organic corn can be a good niche product.

Winter squash is a classic fall vegetable, but again, it takes a lot of space and time to grow in the garden. The plant puts out such large branches that it is best maintained in beds wider than 30 inches.

Many customers love celery (as do we!) but we have not yet been able to produce a crop that satisfies. Growing long, crunchy stalks of celery that measure up to people's expectations is a skill we have not yet mastered. This is one challenge we plan to tackle in the future.

Asparagus is another favorite, but we have no market when it is in season.

Finally, we should mention that we also grow ground cherries, cherry tomatoes, basil, celery root, fennel, Brussels sprouts, and rutabaga, all of which are profitable in the context of small-scale market gardening.

Appendix 2
Tools and Suppliers

Here is a list of market gardening tools and equipment that we use on our micro-farm, along with their commercial names and the suppliers that sell them. Many of these items are not available in traditional gardening stores.

Stirrup Hoe

The stirrup hoes that we use have a sharp, square, swivelling blade. They are made in Switzerland and come in three sizes: 3¼", 5", and 7". Sold by Johnny's Selected Seeds of Maine, USA.

877-564-6697   **johnnyseeds.com**

The 7" blade collinear hoe we use for hoeing salads and leafy vegetables was designed by Eliot Coleman and is also sold by Johnny's Selected Seeds of Maine.

johnnyseeds.com

Bed Preparation Rake

The 30-inch rake that we use to prepare our permanent beds is sold by Johnny's Selected Seeds of Maine, USA.

877-564-6697   **johnnyseeds.com**

Broadfork (Grelinette)

The broadfork that we use is manufactured in Quebec by Denis Bergeron. The tool is 24" wide and has 6 tines. This model has been designed ac-cording to my own personal recommendations in order to make it the most efficient broadfork on the market. Available online from Dubois Agrinovation in Saint-Rémi, Quebec.

800-667-6279   **duboisag.com**

Denis Bergeron has also built us a custom 3-hole punching dibbler that we use for the intensive spacing of ours leeks.

atelierlalibertad.com

Wheel Hoe

A durable, sturdy, and high-quality wheel hoe made by Glaser. The tool has exchangeable blades of different widths: 5", 6", 7", or 8". Sold by Johnny's Selected Seeds of Maine, USA.

877-564-6697   **johnnyseeds.com**

We also use and recommend wheel hoes from the American company Hoss.

hosstools.com

Both the Hoss and Glaser wheel hoes are worth every penny.

Two-Wheel Tractor

The most widely distributed two-wheel tractors in North America are made by the Italian company BCS. I recommend Model 853, as most 30-inch tools can be installed on it. BCS two-wheel tractors are equipped with a rototiller, but there is a wide variety of attachments available: flail mowers, chipper/shredder, plow, ridger, snow thrower, etc. There are various small companies in Italy that make more specialized market gardening tools such as rotary plows, spading machines, and the rotary power harrow we use on our farm. You can get all of these from Earth Tools Inc. of Kentucky, USA.

502-484-3988   **earthtoolsbcs.com**

A Canadian BCS dealer that I have dealt with and recommend is E & F Sauder Sales & Service of Wallenstein, Ontario.

efsaudersales.ca

Greenhouses and Tunnels

We have dealt with a number of companies that make quality greenhouses and tunnels.

Hol-Ser, Inc. of Sainte-Cécile-de-Milton, Quebec.

877-378-6465   **hol-ser.com**

Les Serres Guy Tessier of Saint-Damase, Quebec. Tessier also makes very well-built and affordable caterpillar tunnels.

450-797-3616   **serres-guytessier.com**

Multi Shelter Solutions, from Ontario, offers simple and economical tunnels. Multi Shelter Solutions, Palmerston, Ontario.

866-838-6729   **sheltersolutions.ca**

Irrigation Equipment

Dubois Agrinovation in Saint-Rémi has very competent consultants who are available to help design irrigation plans. We benefited greatly from their expertise when we were designing our system. This company supplied our Naan low-flow sprinklers and Dan micro-sprinklers, as well as all pipes, cam-lock joints, and drip irrigation. Dubois Agrinovation of Saint-Rémi, Quebec.

800-667-6279   **duboisag.com**

Groupe Horticole Ledoux also has many interesting supplies for greenhouse production including tomato clips and hooks, substrate, geotextiles, heat mats, etc. Their catalogue is worth a look. Groupe Horticole Ledoux, Quebec.

888-791-2223   **ghlinc.com**

Indoor Seedling Equipment

We buy all of our cell flats, trays, and pots from Dubois Agrinovation of Saint-Rémi, Quebec.

800-667-6279   **duboisag.com**

Our Precision Vacuum Seeder was homemade, but Johnny's Selected Seeds now sells one that I would recomend. It's quality made and worth its price tage. USA.

877-564-6697   **johnnyseeds.com**

The same company also sells really nice Hydrofarm commercial heat mats that can be chained together to cover whole seedling tables. Johnny's Selected Seeds of Maine, USA.

877-564-6697   **johnnyseeds.com**

Harvest Equipment

Two knives that we really like:

The mini-machete-type field knife is used to harvest broccoli, cauliflower, and lettuce. Sold by William Dam Seeds of Ontario, Canada.

905-628-6641   **damseeds.com**

The #10 Opinel knife is a high-quality knife from France. It is light and very pleasant to handle. Available for order online from Lee Valley Tools, Ottawa, Ontario.

leevalley.com

We sharpen our field knives daily. This job is a lot easier with an ultra-fast sharpener called the Speedy Sharp. Although this sharpener wears the knives down more, its speed and ease of use make it an indispensable tool. We also use it to sharpen the blade of our mesclun harvester. Speedy Sharp is sold in all Canadian Tire stores and is available online.

canadiantire.ca

The garden cart we use to transport our harvest bins from the gardens to our storage area is the Model 26 cart from Carts Vermont. The 26"-diameter wheels of the cart have a track width of about 40", allowing the cart to easily straddle our 30" garden beds as we harvest. This tool is durable, ergonomic, and indispensable. Carts Vermont. Lyndonville, VT.

800-732-7417   **cartsvermont.com**

The Model 26 is also sold by Johnny's Selected Seeds of Maine, USA.

877-564-6697   **johnnyseeds.com**

For mesclun production, we use the Greens Harvester developed by Jonathan Dysingerin of Farmers Friend LCC, in combination with the Six-Row Seeder. As the company says, one person working alone can efficiently harvest 100 pounds of mesclun mix in less than an hour. The tool is simple, affordable, and highly profitable. Sold by Johnny's Selected Seeds of Maine, USA.

877-564-6697   **johnnyseeds.com**

Seeders

All of the seeders used in our gardens (Glaser, EarthWay, and Six-Row) are sold by Johnny's Selected Seeds of Maine, USA.

877-564-6697   **johnnyseeds.com**

The seedbed roller we are experimenting with will be sold by Johnny's Selected Seeds of Maine, USA.

877-564-6697   **johnnyseeds.com**

The hand-pushed finger weeder used for cultivating dense stands of salad mix is commercialized as the Grass Stitcher, a perfect companion the 6-row seeder.

grassstitcher.com

Flame Weeder

The flame weeder that we use to kill pre-emergence weeds in our carrot beds comes from a small American company in West Virginia. We work with the five-torch model, which is 30" wide. Flame Weeders of West Virginia, USA.

304-462-7606   **flame-weeders.com**

Row Cover and Anti-Insect Netting

The floating row cover we use most often is Agryl P19. The anti-insect netting we use for protecting against leek moth has a mesh size of 0.05 inches and is cut to size based on our needs. Both are sold by Dubois Agrinovation of Saint-Rémi, Quebec.

800-667-6279   **duboisag.com**

Dubois also sells the biodegradable and compostable black mulch film BIOTELLO. Dubois Agrinovation of Saint-Rémi, Quebec.

800-667-6279   **duboisag.com**

Backpack Sprayer

Not all sprayers are created equal, and it is worthwhile investing in a high-quality one. If you are trying to cover a length of 330', an efficient sprayer can cut the number of times you need to pump in half. We have been using the same Solo sprayer for more than 10 years. It was originally purchased from Johnny's Selected Seeds.

johnnyseeds.com

The next sprayer we buy will be a more expensive model from Birchmeier.

birchmeier.com

Cold Room

We purchased our cold room second-hand with a new compressor under warranty. The ideal situation may be to deal with a company that sells and repairs air conditioning systems in your area, as after-sales service is very important if your equipment breaks down or is damaged.

An alternative to cooling with a compressor is the CoolBot, which transforms any air conditioner unit (the kind installed in windows) into a cooler that keeps your cold room at a controlled temperature, just as an air compressor would. The CoolBot unit costs a fraction of the price, is easy to install, and consumes less electricity than a conventional compressor. We have not tested it ourselves, but the unit was designed by farmers and comes with a satisfaction guarantee.

888-871-5723   **storeitcold.com**

Seeds

We order most of our seeds from the following companies:

Johnny's Selected Seeds of Maine, USA.

877-564-669   **johnnyseeds.com**

William Dam Seeds of Ontario, Canada.

905-628-6641   **damseeds.com**

West Coast Seeds of Vancouver, Canada.

888-804-8820   **westcoastseeds.com**

High Mowing Organic Seed of Vermont, USA.

802-472-6174   **highmowingseeds.com**

We also import some seeds from les Graines Baumaux in France.

graines-baumaux.fr

Finally, we proudly support the Kokopelli association of France in their fight against Monsanto. The Kokopelli collection offers more than 2,200 varieties of organic heirloom seeds.

kokopelli-semences.fr

Greenhouse Vegetable Seeds

Groupe Horticole Ledoux. 785 Paul-Lussier St., Sainte-Hélène-de-Bagot, Quebec, J0H 1M0. 888-791-2223.

Plant Products. 3370 Le Corbusier Boulevard, Laval, Quebec, H7L 4S8, Canada.

800-361-9184   **plantprod.com**

Deer Fence

You can purchase the different components to build a vertical electric fence at most farm co-ops or feed stores. On a technical level, this involves linking a few "polytapes" (or a number of metal wires with a diameter of 1/10", or 12.5 AWG) on posts spaced every 33'. The wire must be attached to the post by a plastic insulator, and the electrical current circulating must have a tension of at least 4,000 volts. (More is ideal; 8,000 volts will prevent deer from knocking the fence over at night.)

Polypropylene trellises designed expressly for guarding against the entry of deer, raccoons, hares, and other wild animals are sold by Dubois Agrinovation of Saint-Rémi, Quebec.

800-667-6279   **duboisag.com**

Conversion to Vegetable Oil

Our delivery truck and our family vehicle have both been modified to run on recovered vegetable oil. The conversion is possible if you have a diesel motor. The company Greasecar sells conversion kits with comprehensive diagrams and installation instructions. Greasecar Vegetable Fuel Systems, P.O. Box 60508, Florence, Massachusetts, 01062, USA.

413-534-0013 **greasecar.com**

Appendix 3
Garden Plan

Making a garden plan to determine the exact placement of your crops in advance is an important part of crop planning (see Chapter 13 on crop planning). Recall that our farmland is divided into 10 plots of 16 beds organized in families or groups of vegetables. Here is an example of one of our garden plans.

DS = direct-seed   T = transplant   H = harvest

PLOT 1: EARLY CUCURBITACEAE AND BRASSICACEAE

Broccoli	**T:** May 15	
Broccoli	**T:** May 15	
Broccoli	**T:** May 15	
Bok choy	**T:** May 9	
Kohlrabi	**T:** May 9	
Kale	**T:** May 9	
Broccoli	**T:** May 20	Oat/pea green manure seeded in early August; mowed and turned under in late October.
Broccoli	**T:** May 20	
Broccoli	**T:** May 20	
Broccoli	**T:** May 20	
Broccoli	**T:** May 20	
Summer squash	**T:** May 18	
Summer squash	**T:** May 18	
Summer squash	**T:** May 18	
Cauliflower	**T:** June 1	
Cauliflower	**T:** June 1	

PLOT 2: GREENS AND ROOTS

Mesclun mix	**DS:** April 15–June 10	Beet	**DS:** June 30–end of season
Mesclun mix	**DS:** April 15–June 10	Beet	**DS:** June 30–end of season
Snow/snap peas	**DS:** April 19–July 15	Mesclun mix	**DS:** July 27–September 10
Snow/snap peas	**DS:** April 19–July 15	Mesclun mix	**DS:** July 27–September 10
Snow/snap peas	**DS:** April 19–July 15	Mesclun mix	**DS:** July 27–September 10
Snow/snap peas	**DS:** April 19–July 15	Mesclun mix	**DS:** July 27–September 10
Carrot	**DS:** April 20–August 1	Mesclun mix	**DS:** August 10–September 25
Carrot	**DS:** April 20–August 1	Mesclun mix	**DS:** August 10–September 25
Beet	**DS:** April 20–August 1	Mesclun mix	**DS:** August 10–September 25
Turnip	**DS:** April 22–June 20	Mesclun mix	**DS:** August 10–September 25
Spinach	**T:** April 22–June 22	Carrot	**DS:** June 23–end of season
Mesclun mix	**DS:** April 22–June 20	Carrot	**DS:** June 23–end of season
Mesclun mix	**DS:** April 22–June 20	Carrot	**DS:** June 23–end of season
Spinach	**T:** May 16–July 10	Mesclun mix	**DS:** July 13–September 1
Spinach	**T:** May 16–July 10	Mesclun mix	**DS:** July 13–September 1
Spinach	**T:** May 16–July 10	Mesclun mix	**DS:** July 13–September 1

PLOT 3: GARLIC

Garlic	**DS:** October; **H:** July of the following year	
Garlic		
Garlic		
Garlic		
Garlic		
Garlic		
Garlic		
Garlic		Oat/pea green manure seeded in early August; mowed and turned under in late October.
Garlic		
Garlic		
Garlic		
Garlic		
Garlic		
Garlic		
Garlic		
Garlic		

PLOT 4: GREENS AND ROOTS

Vetch/oat green manure seeded April 15; turned under in early June.	Mesclun mix	**DS:** June 29–August 15	Garlic	**DS:** October 15
	Mesclun mix	**DS:** June 29–August 15	Garlic	**DS:** October 15
	Mesclun mix	**DS:** June 29–August 15	Garlic	**DS:** October 15
	Mesclun mix	**DS:** June 29–August 15	Garlic	**DS:** October 15
	Bean	**DS:** June 21–September 1	Garlic	**DS:** October 15
	Bean	**DS:** June 21–September 1	Garlic	**DS:** October 15
	Bean	**DS:** June 21–September 1	Garlic	**DS:** October 15
	Bean	**DS:** June 21–September 1	Garlic	**DS:** October 15
	Carrot	**DS:** June 8–October 1	Garlic	**DS:** October 15
	Carrot	**DS:** June 8–October 1	Garlic	**DS:** October 15
	Carrot	**DS:** June 8–October 1	Garlic	**DS:** October 15
	Carrot	**DS:** June 8–October 1	Garlic	**DS:** October 15
	Lettuce	**T:** June 15–July 15	Garlic	**DS:** October 15
	Lettuce	**T:** June 15–July 15	Garlic	**DS:** October 15
	Lettuce	**T:** June 28–August 15	Garlic	**DS:** October 15
	Lettuce	**T:** June 28–August 15	Garlic	**DS:** October 15

PLOT 5: SOLANACEAE

Eggplant	**T:** May 30–end of season
Eggplant	**T:** May 30–end of season
Eggplant	**T:** May 30–end of season
Eggplant	**T:** May 30–end of season
Eggplant	**T:** May 30–end of season
Ground cherry	**T:** May 30–end of season
Ground cherry	**T:** May 30–end of season
Hot pepper	**T:** May 30–end of season
Pepper	**T:** May 30–end of season
Pepper	**T:** May 30–end of season
Melon	**T:** June 6–end of season
Melon	**T:** June 6–end of season
Melon	**T:** June 6–end of season
Melon	**T:** June 6–end of season
Melon	**T:** June 6–end of season
Melon	**T:** June 6–end of season

PLOT 6: GREENS AND ROOTS

Bean	**DS:** May 23–August 10	Spinach	**T:** August 15–October
Bean	**DS:** May 23–August 10	Spinach	**T:** August 15–October
Radish	**DS:** May 23–July 1	Lettuce	**T:** July 12–August 12
Turnip	**DS:** May 23–July 10	Lettuce	**T:** July 12–August 12
Carrot	**DS:** May 25–August 20	Arugula	**DS:** August 25–October
Carrot	**DS:** May 25–August 20	Mustard greens	**DS:** August 25–October
Arugula	**DS:** May 20–July 5	Swiss chard	**T:** July 10–October
Arugula	**DS:** May 20–July 5	Swiss chard	**T:** July 10–October
Lettuce	**T:** May 31–July 1	Bean	**DS:** July 20–mid-September
Lettuce	**T:** May 31–July 1	Bean	**DS:** July 20–mid-September
Beet	**DS:** June 6–August 25	Arugula	**DS:** September 1–end of season
Beet	**DS:** June 6–August 25	Arugula	**DS:** September 1–end of season
Bean	**DS:** June 6–August 20	Spinach	**T:** August 25–end of season
Bean	**DS:** June 6–August 20	Spinach	**T:** August 25–end of season
Bean	**DS:** June 6–August 20	Spinach	**T:** August 25–end of season
Bean	**DS:** June 6–August 20	Spinach	**T:** August 25–end of season

PLOT 7: SUMMER CUCURBITACEAE AND BRASSICACEAE

	Broccoli	**T:** June 25
	Broccoli	**T:** June 25
	Broccoli	**T:** June 25
	Broccoli	**T:** June 25
	Cabbage	**T:** June 25
	Cabbage	**T:** June 25
	Brussels sprouts	**T:** June 25
Vetch/oat green manure seeded April 15; turned under in early June.	Summer squash	**T:** July 5
	Cauliflower	**T:** July 14
	Cauliflower	**T:** July 14
	Broccoli	**T:** July 14
	Broccoli	**T:** July 14
	Broccoli	**T:** July 14
	Broccoli	**T:** July 14
	Kohlrabi	**T:** July 27
	Kohlrabi	**T:** July 27

PLOTS 8: GREENS AND ROOTS

Carrot	DS: May 4–July 25	Lettuce	T: July 27–August 27
Carrot	DS: May 4–August 4	Lettuce	T: July 27–August 27
Carrot	DS: May 4–August 4	Lettuce	T: August 12–September 12
Carrot	DS: May 4–August 4	Lettuce	T: August 12–September 12
Cilantro/Dill	DS: May 15–August 4	Mesclun mix	DS: August 24–October
Turnip	DS: May 6–July 1	Mesclun mix	DS: August 24–October
Radish	DS: May 10–July 1	Mesclun mix	DS: August 24–October
Arugula	DS: May 10–July 1	Mesclun mix	DS: August 24–October
Beet	DS: May 10–August 10	Mesclun mix	DS: September 7–end of season
Beet	T: May 16–August 10	Mesclun mix	DS: September 7–end of season
Chicory	T: May 11–August 10	Mesclun mix	DS: September 7–end of season
Snow/snap peas	DS: May 13–August 10	Mesclun mix	DS: September 7–end of season
Snow/snap peas	DS: May 13–August 10	Mesclun mix	DS: September 7–end of season
Snow/snap peas	DS: May 13–August 10	Chinese cabbage	T: August 15–end of season
Snow/snap peas	DS: May 13–August 10	Collard greens	T: August 15–end of season
Lettuce	T: May 16–June 25	Fennel	T: July 28–end of season

PLOT 9: LILIACEAE

Green onion	**T:** May 1	
Green onion	**T:** May 1	
Green onion	**T:** May 1	
Green onion	**T:** May 1	
Green onion	**T:** May 1	
Green onion	**T:** May 1	
Leek	**T:** May 5	
Leek	**T:** May 5	Oat/pea green manure seeded in early September and left as ground cover for the winter.
Leek	**T:** May 5	
Leek	**T:** May 5	
Storage onion	**T:** May 8	
Storage onion	**T:** May 8	
Storage onion	**T:** May 8	
Storage onion	**T:** May 8	
Storage onion	**T:** May 8	
Storage onion	**T:** May 8	

PLOT 10: GREENS AND ROOTS

Mesclun mix	**DS:** May 4–June 20	Carrot	**DS:** July 5–end of season
Mesclun mix	**DS:** May 4–June 20	Carrot	**DS:** July 5–end of season
Mesclun mix	**DS:** May 4–June 20	Bean	**DS:** July 4–end of season
Mesclun mix	**DS:** May 4–June 20	Bean	**DS:** July 4–end of season
Mesclun mix	**DS:** May 18–July 5	Winter radish	**DS:** July 11–end of season
Mesclun mix	**DS:** May 18–July 5	Winter radish	**DS:** July 11–end of season
Mesclun mix	**DS:** May 18–July 5	Winter radish	**DS:** July 11–end of season
Mesclun mix	**DS:** May 18–July 5	Chicory	**T:** July 11–end of season
Mesclun mix	**DS:** June 1–July 15	Kale	**T:** August 5–end of season
Mesclun mix	**DS:** June 1–July 15	Kale	**T:** August 5–end of season
Mesclun mix	**DS:** June 1–July 15	Chinese cabbage	**T:** August 8–end of season
Mesclun mix	**DS:** June 1–July 15	Chinese cabbage	**T:** August 8–end of season
Mesclun mix	**DS:** June 15–August 1	Parsley	**T:** August 5–end of season
Mesclun mix	**DS:** June 15–August 1	Turnip	**DS:** August 5–October
Mesclun mix	**DS:** June 15–August 1	Radish	**DS:** August 20–October
Mesclun mix	**DS:** June 15–August 1	Turnip	**DS:** August 25–end of season

TUNNEL 1

Mesclun mix	**DS:** March 5–April 10
Mesclun mix	**DS:** March 5–April 10
Mesclun mix	**DS:** March 5–April 10
Mesclun mix	**DS:** March 10–April 20
Mesclun mix	**DS:** March 10–April 20
Beet	**T:** April 20–July 1
Carrot	**DS:** April 20–July 1
Summer squash	**T:** April 24–July 15
Cucumber	**T:** April 25
Cucumber	**T:** April 25
Cucumber	**T:** July 25
Cucumber	**T:** July 25
Mesclun mix	**DS:** September 25
Mesclun mix	**DS:** September 25
Mesclun mix	**DS:** September 25

TUNNEL 2

Mesclun mix	**DS:** March 20–April 27
Mesclun mix	**DS:** March 20–April 27
Mesclun mix	**DS:** March 20–April 27
Mesclun mix	**DS:** March 28–May 5
Mesclun mix	**DS:** March 28–May 5
Pepper	**T:** May 1
Pepper	**T:** May 1
Pepper	**T:** May 1
Cucumber	**T:** June 15
Cucumber	**T:** June 15
Mesclun mix	**DS:** October 5
Mesclun mix	**DS:** October 5
Mesclun mix	**DS:** October 5
Mesclun mix	**DS:** October 10
Mesclun mix	**DS:** October 10

GREENHOUSE

	Tomato	**T:** April 16
	Tomato	**T:** April 16
	Tomato	**T:** April 16
Plant Nursery	Tomato	**T:** May 20
	Tomato	**T:** May 20
	Tomato	**T:** May 20
	Basil	**T:** May 20

Appendix 4
Annotated Bibliography

Below are some of the works I used to write *The Market Gardener*, as well as others I find useful in setting up and operating a vegetable micro-farm. Most of these documents were written with large-scale vegetable growing or organic gardening in mind, but for market gardeners ideas from either scale of production can be useful.

Adabio. *Guide de l'autoconstruction : Outils pour le maraîchage biologique.* France: Édition Adabio-Itab, 2012. Adabio is a coalition of French organic farmers who share their agricultural machine designs in a DIY and open source fashion. This instruction manual is geared toward tractor implements, but the merit and quality of the book makes it well worth a look. In French.

Allen, Will and Charles Wilson. *The Good Food Revolution: Growing Healthy Food, People, and Communities.* New York: Gotham books, 2012. Will Allen is the founder of Growing Power, an innovative urban farm located in downtown Milwaukee. This book tells his story.

Altieri, Miguel A., Clara I. Nicholls, and Marlene A. Fritz. *Manage Insects on Your Farm: A Guide to Ecological Strategies.* Belstville, MD: Sustainable Agriculture Network, 2005. This book explains how to set up your farm to mitigate the impact of certain insect pests. Although devised for California-type weather, the principles of ecological pest management are universal.

Asselineau, Eléa and Gilles Domenech. *Les bois raméaux fragmentés. De l'arbre au sol.* Arles, France: Éditions du Rouergue, 2007. Ramial chipped wood (RCW) is a specific soil regeneration technique invented and developed in Quebec. This book explains the why and the how of it, including the basics of adding forest materials onto vegetable production systems. It's the only book I know on the subject. In French.

Blanchard, Chris and Paul Dietmann, and Craig Chase. *Fearless Farm Finances: Farm Financial Management Demystified.* Spring Valley, WI: Midwest Organic and Sustainable Education Service, 2012. This book will help you understand basic financial management on your farm—how to collect and use numbers in ways that will give you real information that is helpful in making farm decisions.

Bradbury, Zoe Ida, Severine von Tscharner Fleming, and Paula Manalo. *Greenhorns: 50 Dispatches from the New Farmers' Movement.* North Adams, MA: Storey Publishing LLC, 2012. This is a feel good book about who we are as movement. Reading it will most certainly make you want jump on the bandwagon.

Bourguignon, Claude and Lydia. *Le sol, la terre et les champs : Pour retrouver une agriculture saine*. Paris, France: Éditions Sang de la Terre, 2008. The two authors are among the rare soil microbiologists researching the effects of conventional agriculture on organisms living underground. This book helps you to understand how the soil's fauna plays a key role in soil fertility and how fragile this system can be. A leading document in French agroecology. In French.

Byczynski, Lynn. *Market Farming Success, Revised and Expanded Edition: The Business of Growing and Selling Local Food*. White River Junction, VT: Chelsea Green Publishing, 2013. The author is the editor of *Growing for Market*, a market gardening periodical published since 1992. Byczynski knows her stuff, and in this book she discusses, among other topics, the differences between market gardens, market farms, and vegetable farms and the income that can be expected from each scale of production. A must read for the aspiring market gardener.

Caldwell, Brian, Eric Sideman, Abby Seaman, Anthony Shelton, and Christine D. Smart. *Resource Guide for Organic Insect and Disease Management*, 2nd ed. Geneva, NY: New York State Agricultural Experiment Station, 2005. Currently one of the few reference guides that describe biopesticides (where they come from and what effects they have) along with diseases and insect pests. Available for download from: web.pppmb.cals.cornell.edu/resourceguide.

Clark, Andy, editor. *Managing Cover Crops Profitably*, 3rd ed. College Park, MD: SARE Outreach, 2012. One of the most comprehensive and useful documents on green manures. Updated regularly and available for free online from the organization's website: sare.org.

Coleman, Eliot. *The New Organic Grower: A Master's Manual of Tools and Techniques for the Home and Market Gardener*, 2nd ed. White River Junction, VT: Chelsea Green Publishing, 1995. This classic is the first book on vegetable growing that I read and remains the most influential. Although the technical information is sometimes incomplete, the book is still a good introduction to growing on a small scale. Written by a grower, for growers.

Coleman, Eliot. *The Winter Harvest Handbook: Year-Round Vegetable Production Using Deep Organic Techniques and Unheated Greenhouses*. White River Junction, VT: Chelsea Green Publishing, 2009. Another one of my favorite books, this work is a testament to Coleman's 40 years of horticultural experience and innovation. This is an invaluable resource for anyone interested in extending his or her growing season.

Edey, Anna. *Solviva: How to Grow $500,000 on One Acre and Peace on Earth*. Vineyard Haven, MA: Trailblazer Press, 1998. A little-known but very interesting book that discusses the idea of integrating a salad farming operation into a solar home and greenhouse. Filled with great ideas.

Ellis, Barbara W. and Fern Marshall Bradley, editors. *The Organic Gardener's Handbook of Natural Insect and Disease Control. A Complete Problem-Solving Guide to Keeping Your Garden and Yard Healthy Without Chemicals*. Emmaus, PA: Rodale Books, 1996. One of the better books out there about biocontrol of insect pests and diseases.

Elizabeth Henderson and Robyn Van En. *Sharing the Harvest: A Citizen's Guide to Community Supported Agriculture*. White River Junction, VT: Chelsea Green Publishing, 2007. A very comprehensive look at the CSA model. Farmers will find lots of useful advice on structuring and organizing a CSA share program.

Falk, Ben. *The Resilient Farm and Homestead: An Innovative Permaculture and Whole Systems Design Approach*. White River Junction, VT: Chelsea Green Publishing, 2013. One of the better permaculture books recently published. Although written for the homestead, this book gives great insights into how technical (and worthwhile) the design stage of setting up a farm can get.

Fukuoka, Masanobu. *The Natural Way of Farming: The Theory and Practice of Green Philosophy*. Mapusa, Goa, India: Other India Press, 1985. Fukuoka's works are seminal texts for those interested in permaculture. The author's approach is quite radical..

Henderson, Elizabeth and Karl North. *Whole Farm Planning: Ecological Imperatives, Personal Values, and Economics*. White River Junction, VT: Chelsea Green Publishing, 2004. This book draws on the theory of holistic management, adapting it to the context of diversified market farming. Useful for understanding the importance of setting financial objectives (and non-financial objectives) at the very beginning of crop planning.

Holzer, Sepp. *Sepp Holzer's Permaculture: A Practical Guide to Small-Scale, Integrative Farming and Gardening*. White River Junction, VT: Chelsea Green Publishing, 2011. Sepp Holzer is a living legend in permaculture circles. The ideas that he proposes are based on his own experience, which is, alas, rare among those who write about permaculture.

Hopkins, Rob. *The Transition Handbook: From Oil Dependency to Local Resilience*. Cambridge, UK: Green Books, 2008. This book explains how to organize our communities for the end of cheap oil. It focuses not on the catastrophe, but instead on positive changes we must make at the local level. A must-read.

Howard, Ronald J., J. Allan Garland, and W. Lloyd Seaman, editors. *Diseases and Pests of Vegetable Crops in Canada*. Ottawa, ON: The Entomological Society of Canada, 2007. The best reference book for identifying insect pests and crop diseases in Canada.

Hunt, Marjorie B. and Brenda Bortz. *High-Yield Gardening: How to Get More from Your Garden Space and More from Your Gardening Season*. Emmaus, PA: Rodale, 1986. One of the first resource books we consulted on intensive growing techniques. The methods described are for a non-commercial scale.

Jeavons, John. *How to Grow More Vegetables and Fruits, Nuts, Berries, Grains and Other Crops Than You Ever Thought Possible on Less Land Than You Can Imagine*, 8th edition. Berkeley, CA: Ten Speed Press, 2012. This celebrated work is well worth the read even though much emphasis is placed on double-digging, which is not essential in my view.

Kimball, Kristin. *The Dirty Life: A Memoir of Farming, Food and Love*. New York, NY: Scribner, 2010. Essex Farm is one of the most interesting farms I have visited. In this book, the author, who co-owns the business, recounts the beginnings of the farm's CSA project. It provides a good description of the harsh reality of the first years of establishing a farm along with some lessons that can be drawn, and the payoff after many years of hard dirty work.

Kuroda, Tatsuo. *EM : Les micro-organismes efficaces pour le jardin*. Paris, France: Le Courrier du Livre, 2010. This is not the best book on gardening, but it is one of the few currently available that gives recommended doses of effective micro-organisms (EM) in the garden, a topic that interests me. In French.

Dawling, Pam. *Sustainable Market Farming: Intensive Vegetable Production on a Few Acres*. New Society Publishers, 2013. A reference book like Dawling's was long in the waiting. Intended for serious growers, it gives a comprehensive overview of growing techniques and practices for almost every vegetable. A good addition to any market gardener's library.

Matson, Tim. *Earth Ponds: The Country Pond Maker's Guide to Building, Maintenance, and Restoration*, 3rd ed. Woodstock, VT: Countryman Press, 2012. This guide contains all the considerations and steps involved in creating a lake or pond to serve as an ecological habitat. This book is well-written and penned by a foremost authority on the topic.

Lowenfels, Jeff and Wayne Lewis. *Teeming with Microbes: The Organic Gardener's Guide to the Soil Food Web*. Portland, OR: Timber Press, 2010. A wonderful book that clearly describes the living systems contained within the soil. Helps us understand the perverse effects of inverting soil and how biology can replace mechanical tillage. A must-read.

Magdoff Fred and Harold Van Es. *Building Soils for Better Crops: Sustainable Soil Management*. College Park, MD: SARE Outreach, 2009. This book describes the relationship between soil biology and organic crop fertilization by focusing on organic matter management. Written in an engaging and easy-to-read style.

Mollison, Bill and David Holmgren. *Permaculture One: A Perennial Agriculture for Human Settlements*. Sisters Creek, Tasmania, Australia: Tagari, 1981. This book is a veritable encyclopedia of permaculture. Even though the ideas presented are more relevant for a subtropical climate, the concepts are universal.

Moreau, J. G. and J. J. Daverne. *Manuel pratique de la culture maraîchère de Paris*. Paris, France: Imprimerie Bouchard-Huzard, 1845. An extraordinary reference document detailing the growing practices of French market gardeners of the 19th century. Reading about their methods makes you realise how these growers were amazingly productive. Out of print, but available through Abebooks.com. In French.

Nearing, Helen and Scott. *Living the Good Life: How to Live Sanely and Simply in a Troubled World*. New York: Schocken Books, 1973. This book is an American back-to-the-land classic. It recounts the adventures of an eminent communist professor and his young theosophist wife, who moved out into the country in the 1930s to lead a self-sufficient life. Their life philosophy, presented in the context of the era, makes for a unique read.

Raymond, Hélène and Jacques Mathé. *Une agriculture qui goûte autrement : Histoires de productions locales de l'Amérique du Nord à l'Europe*.

Québec, QC: Éditions MultiMondes, 2011. This collection of inspiring farming stories endeavors to map the small-scale agriculture movement that is currently emerging in Europe and the Americas. In French.

Schwarz, Michiel and Diana Krabbendam. *Sustainist Design Guide: How Sharing, Localism, Connectedness and Proportionality Are Creating a New Agenda for Social Design*. Amsterdam, NL: BIS Publishers, 2013. This book doesn't talk about farming, nor about vegetable growing, but it taps into other skill sets that are vital for successful market gardening.

Stamets, Paul. *Mycelium Running: How Mushrooms Can Help Save the World*. Berkeley, CA: Ten Speed Press, 2005. A great book, making a major contribution towards our understanding of the importance of mycelium in the creation of soil.

Thériault, Frédéric and Daniel Brisebois. *Crop Planning for Organic Vegetable Growers*. Ottawa, ON: Canadian Organic Growers, 2010. This is the best book on crop planning. It uses the example of a young couple starting a farm to present its method—a very practical approach.

Tickell, Joshua. *From the Fryer to the Fuel Tank: The Complete Guide to Using Vegetable Oil as an Alternative Fuel*. New Orleans, LA: Joshua Tickell Publications, 2003. For 10 years now, we have been powering our vehicles with recycled vegetable oil. The author covers the topic thoroughly in this book and shows the reader how to convert a diesel vehicle, step by step.

Tompkins, Peter and Christopher Bird. *Secrets of the Soil: New Solutions for Restoring Our Planet*. Anchorage, AK: Earthpulse Press, 1998. Although admittedly esoteric, this is one of my favorite books. It helps us to imagine what discoveries could be found if science got interested ecological agriculture. A fascinating work.

Walters, Charles. *Eco-Farm: An Acres U.S.A. Primer*, 3rd ed. Austin, TX: Acres U.S.A., 2003. This was written by the founder of Acres U.S.A., one of the first organizations to make the case for an ecological approach to agriculture from a scientific perspective. While it is not an easy read, it does allow the reader to deepen his or her understanding of soil with respect to fertilization.

Wiediger, Paul and Alison. *Walking to Spring: Using High Tunnels to Grow Produce 52 Weeks a Year*. Smiths Grove, KY: Au Naturel Farm, 2003. Market farmers from Kentucky, the Wiedigers share much of their learned experience in this self-published guide. A good primer for growing in hoophouses even though written for warmer climates than that of the Northeast.

Wiswall, Richard. *The Organic Farmer's Business Handbook: A Complete Guide to Managing Finances, Crop, and Staff—and Making a Profit*. White River Junction, VT: Chelsea Green Publishing, 2009. Written by a veteran grower, this book discusses the financial aspect of running a CSA vegetable farm. The chapter on planning for retirement is especially interesting.

Selected Organizations and Their Websites

ACORN (Atlantic Canadian Organic Regional Network) is eastern Canada's flagship organization for the organic agricultural community and offers an apprenticeship and mentorship program for beginning farmers. ACORN's website is full of interesting information for both small and large-scale organic growers, and their annual ACORN conference is well worth the trip. acornorganic.org

CAPE (Coopérative pour l'Agriculture de Proximité Ecologique) gathers Quebec's ecological farmers to promote and develop the market farming sector. CAPE organizes conferences, workshops, farm visits, and helps the creation of new proximity markets through collective marketing, group purchasing, and lobbying. capecoop.org

CETAB+ (Centre d'expertise et de transfert en agriculture biologique et de proximité) leads a number of applied research projects related to organic and local agriculture. Its staff includes some of the most respected experts in organic agriculture in Quebec. Their website features lots of information and has a free database that reviews all technological and scientific advances in organic agriculture in French. This resource is a goldmine. cetab.org

COG (Canadian Organic Growers) is a national organization that publishes an excellent quarterly magazine filled with interesting articles written by small growers. These articles are always very well done. Becoming a member of COG gives you access to their free mail book borrowing system. Their website is also worth a visit. cog.ca

Greenhorns is a grassroots network of organizers, artists, and farmers whose mission is to recruit, promote, and support the new generation of young farmers in the US. They do this by producing avant-garde programming and publications, and hosting events enhancing the social and cultural lives of young farmers. Greenhorns is a founding partner in many spin-offs including National Young Farmers Coalition, Farm Hack, and Agrarian Trust. thegreenhorns.net

Équiterre coordinates the largest network of CSA farms in Canada. The organization publishes information resources, organizes farm visits, and offers many other activities. Their support for small farm start-ups is a concrete and very useful tool. equiterre.org

FarmStart is a charitable organization in Canada that provides tools, resources, and support to help a new generation of entrepreneurial, ecological farmers to get their farms off the ground and to thrive. They do this by connecting new farmers with training resources and mentors. farmstart.ca

Growing for Market. In my opinion, this monthly magazine is by far the most useful resource for market gardeners. The articles are written by growers who share their production methods, tips, and recommendations. The magazine is available online as well. growingformarket.com

Young Agrarians is a grassroots initiative from British Columbia, focused on connecting and recruiting the next generation of young farmers through mixers, a blog, a farmer resource map centralizing information about sustainable agriculture, and more. Check them out and join the movement here: youngagrarians.org

Appendix 5
Glossary

Alternaria: Fungal disease that infects leaves and can cause dieback in plants. It is common in tomatoes and appears as brown spots in concentric circles.

Amend: To incorporate substances (e.g., organic matter, clay, lime) into the soil to improve its physical and biological fertility, unlike fertilizers, which improve chemical fertility.

Basalt: Black volcanic rock, used as a fertilizer in powder form.

Bed: Growing in beds is the technique of organizing the garden into rows separated by pathways. The bed width is determined in advance and measured from the middle of one pathway to the middle of the next.

Bioactivators: Different preparations of micro-organisms (mycorrhizae, bacteria, etc.) that improve soil fertility by increasing the availability of nutrients already present in the soil. "Biostimulant" is an equivalent term.

Biocontrol agents: Living organisms, generally insects, used to control crop pests. For example, trichogramma are tiny wasps that parasitize the European corn borer at the larval (caterpillar) stage.

Biodynamics: An approach to agriculture proposed in 1924 by the anthroposophist Rudolph Steiner (1861–1925). The main agricultural practices that characterize biodynamics are the application of biodynamic preparations (e.g., cow horn manure) to compost and plants to stimulate beneficial interactions (e.g., those involving micro-organisms), the use of a calendar to plan operations based on lunar phases and constellations, and having a complete life cycle on the farm that includes growing plants and raising animals.

Biopesticides: Plant protection products made of plant extracts (e.g., pyrethrum) or micro-organisms or their derivatives (e.g., *Bacillus thuringiensis* or Bt). These products are generally available in liquid or wettable powder form.

Blossom-end rot: A physiological disease that is common in peppers and tomatoes and that generally occurs when the weather changes from dry to wet. Blossom-end rot is caused by a lack of calcium in the soil or irregular watering, which limits the availability of calcium to the plant. It appears as a circular black spot that develops at the base of the fruit. It is common to see opportunistic fungi colonizing this fragile zone.

Bolting: Process during which a plant goes to seed. The harvest may be lost if the plants bolt prematurely, which is usually caused by extreme climate conditions.

Brassicaceae: A botanical family that includes many vegetables such as broccoli, cabbage, turnips, and radish. The plants, also called crucifers, are recognized by their cross-shaped flower (from which the name "crucifer" is derived).

Broadfork: A long U-shaped fork with several teeth that drive into the soil vertically. This ergonomic gardening tool makes it possible to use leverage to work the soil deep down without turning it.

Btk: *Bacillus thuringiensis* var. *kurstaki* (*Btk*) is a species of bacteria that is naturally present in soil. It is used as a biopesticide to suppress the population of many insect pests in agriculture and forestry. In market gardening, it is mainly used to control lepidopterans (butterflies and moths).

Canopy: In horticulture, this term refers to the upper level of the plants where the leaves are found. If the crops are planted closely together, the canopy "umbrella" creates a micro-climate and limits weed growth.

Capillary action: The process by which liquid is pulled up through the soil. Water climbs through capillary action in the micro-pores that are created by compaction of the soil surface. This process is what allows moisture deep down in the soil to reach the plant growth zone near the surface.

Channel: Small, sloped ditch for water drainage.

Chisel plow: A deep tillage tool pulled by a tractor. The tool is equipped with fixed teeth that tear up and loosen the soil without turning it. It was invented to replace the plow, in order to preserve a cover of crop residues on the soil surface to prevent erosion. The tool dates from the Dust Bowl period, when a series of dust storms struck the United States in the 1930s.

Compaction: An increase in soil density caused by packing of the upper layer. In market gardening, the main causes of compaction are the weight of the tools and the number of trips they make as well as excessive tillage.

Cotyledons: The first leaves that appear on seedlings that have just germinated from seed. Note that dicotyledonous plants (e.g., legumes) have two cotyledons while monocotyledonous plants (e.g., grasses) have only one.

Couch grass: Perennial weed in the Poaceae family. It is difficult to control in crops because of its vigorous rhizomes, which propagate quickly and multiply when cut by the action of a hoe or rotary tilling machine.

Crop rotation: The technique of growing different groups of crops in succession on the same plot.

Crop system: Set of procedures adopted by a vegetable grower for his or her production. A variety of practices make up the cropping system: permanent beds, crop rotation, succession, etc.

Cucurbitaceae: A family of plants with creeping, often thick, vines and large fruits (e.g., pumpkin, squash, cucumber, and melon).

Cultivar: A variety of vegetable species developed by humans through genetic selection with the goal of encouraging certain characteristics including beauty, productivity, growth rate, and resistance to certain diseases. The terms "variety" and "cultivar" are used interchangeably by market gardeners.

Cultivating: Scraping the soil superficially, using a hoe or other tool, to eliminate weeds. The terms "hoeing" and "cultivating" are often used interchangeably, as the same tools can be used to do both. However, the objective of hoeing is to aerate the soil, not to weed it.

Damping off: A disease that appears as the base of a plant's stem becoming shriveled and lanky, which leads to weakness and subsequent death of the plant. It is common in plant nurseries, but there are few ways to fight it. This is why preventative measures are important.

Deficiency: A condition that results when a plant lacks a substance essential to its growth. It may be caused by lack of nutrients in the soil or the unavailability of these nutrients. Deficiencies can cause any number of symptoms, notably discoloration in the leaves. They are not to be confused with phytosanitary problems (e.g., pathogenic fungi and bacteria, insect pests, viruses).

Drip irrigation: A technique that uses a polyethylene pipe in which several emitters (also called "drippers") distribute water to the base of the plants in a slow and controlled fashion. This watering system is much more precise and economical (in terms of water usage) than sprinkler systems. It is also called "trickle irrigation" or "micro-irrigation."

Dumping: A commercial practice that consists of selling a product for less than it cost to produce (i.e., at a loss) with the goal of selling the merchandise quickly or beating the competition. The practice is generally considered underhanded.

Early crops: The first vegetables of a crop harvested outside their normal season. Producing early crops can give farmers a major competitive advantage at market. Different techniques (row cover, tunnels, greenhouses, etc.) can be used to grow them.

Energy efficiency: A strategy to save energy through various means. In the case of a greenhouse, energy efficiency is mostly a question of insulating the building, eliminating drafts, and using thermal screens and heat mats.

Erosion: Degradation of the soil by atmospheric elements, chiefly wind and water, leading to the loss of the topsoil, which is the most productive soil in vegetable production. In Quebec, the main factor leading to erosion is rain, which causes runoff at the soil surface if the soil is not held in place by plant roots. Note that erosion is a horizontal motion, while leaching is a vertical motion.

Etiolated: Describes a plant that lacks light and grows taller to compensate. Etiolated plants are leggy, discolored, and less vigorous.

Fertigation: The technique of applying water-soluble fertilizers via an irrigation system. The term is a combination of the words "fertilization" and "irrigation."

Fertilizer: An organic or mineral substance incorporated into the soil to maintain or increase fertility.

Fertilizing waste substances: Name given to all organic or mineral substances from human activity that have fertilizing potential. This includes sludge (also called "biosolids") and compost. Management (e.g., storage and spreading) of fertilizing waste substances is governed by government standards.

Flame weeder: Device that produces flames used to eliminate weeds by thermal shock rather than calcination.

Flea beetle: Member of a group of small beetles with shiny black shells that jump when disturbed. Flea beetle damage is easy to spot; the adults leave small, roundish holes in the leaves.

Forcing: The expression "forcing" crops comes to us from the Parisian market gardeners of the 19th century (*forcer* in French). While the term is less common today, it refers to the use of different horticultural techniques to bring a crop to harvest before the end of its normal growth duration. Forcing makes it possible to grow early vegetables.

Fruiting period: The period during which a plant forms its fruits.

Genetically modified organism (GMO): Organism whose genome has been modified by genetic engineering to give it properties it would not naturally possess. The presence of GMOs in the food system and the environment is the source of much concern.

Greenhouse grower: A vegetable grower specializing in greenhouse production.

Green manure: A crop intended to amend the soil, to prevent erosion and leaching of nutrients, and/or to fight weeds (e.g., by smothering or allelopathy). Green manures are not intended for sale.

Green revolution: The name for the technological leap in agriculture between 1960 and 1990. The green revolution is characterized by increased specialization in crops, which, in conjunction with the use of synthetic chemicals and pesticides, resulted in a spectacular increase in agricultural productivity.

Hardening off: Exposing transplants to harsh weather conditions in order to increase their resistance once they are planted in the garden. The term "acclimatize" is also used.

Hardiness zones: The hardiness scale is a series of codes attributed to ornamental plants based on their capacity to resist frost. The geographic zones are defined based on a formula that takes account of several climatic factors influencing plant hardiness in a given area. The minimum winter temperature is the most important factor with respect to plant survival.

Hilling: Forming a mound ("hill") at the base of a plant.

Hoeing: The action of loosening and aerating the upper layer of soil around the crops.

Horticultural techniques: Methods used in vegetable production. Examples include the false seedbed technique, flame weeding, and pruning tomato plants. Also called "growing techniques."

Inflorescence: A plant organ with a cluster of

flowers such as that found in the centre of a broccoli or cauliflower plant.

Leaf spot: A fungal disease affecting plant leaves, frequently found in carrots and beets. The pathogenic fungus makes a dark brown spot that eventually transforms into localized necrosis (dead, dry tissue). In serious cases, the spots may multiply and kill the leaves of the crop.

Liming: Amending the soil with lime or other calcic substances. This practice is often necessary in soil that is overly acidic or deficient in calcium.

Loam: A type of soil made up of a relatively even mix of sand, silt, and clay particles. Depending on the exact proportions of these particles, the loam may be further described as sandy, silty, or clayey.

Market gardener: An artisan of the soil who does his or her work on a small cultivated area in greenhouses and open fields. He or she produces a wide variety of vegetables and sells them directly to consumers.

Member: Name given to a "customer" of a community-supported agriculture (CSA) farm service. Unlike normal customers, members purchase part of the harvest in advance, thus sharing with the farmer the risk involved in farming.

Microclimate: Special climatic conditions in a localized area (e.g., a valley, a site, a farm) that differ from the rest of the area. Microclimates may be favorable to specific crops (because of factors such as humidity, temperature, and sun).

Microgreens: Young leaves of a crop of greens.

Mineralization: The release of mineral elements (nitrogen, potassium, phosphorus, etc.) contained in organic matter via the action of biological activity in the soil. Mineralization makes available the nutrients needed for plant growth.

Mite: Tiny insect predator used for biocontrol, mainly in greenhouse production.

Node: Part of the plant where a new leaf develops from the main stem.

Organic matter: All living or dead material of animal or plant origin found in the soil. It is in most soil in varying proportions (usually between 0.5% and 10%). Fresh organic matter is made of leaves, twigs, crop residues, roots, micro-organisms, etc. Decomposed organic matter forms the humus in the soil.

Peat moss: Spongy organic material resulting from the slow decomposition of plants (*Sphagnum* mosses) in the wet, acidic, oxygen-poor conditions of bogs.

Peduncle: Point of attachment between the fruit and the stem of the plant.

Permaculture: Farm system design developed in the 1970s by Australians Bill Mollison and David Holmgren. Founded on the principles of ecology and design, permaculture endeavors to create self-managed, productive, and energetically efficient agricultural systems. Today, it applies to all spheres of human activity through initiatives such as Cities in Transition.

Pests: Insects and other organisms (e.g., mammals and birds) that harm crops.

Phenology: The study of the influence of climate on the development of plants (foliation, flowering, fruiting) and animals.

Plant nursery: Place where young plants are grown for the purpose of transplanting, usually a greenhouse.

Plant protection: A set of preventative and curative strategies designed to protect crops from pests in light of their break-even point.

Potassium sulfate: Ground rock used as a natural fertilizer in organic agriculture.

Potting up: Action of transplanting seedlings grown in containers into larger containers in order to give them more space to grow.

Powdery mildew: A disease that is caused by microscopic fungi, which appear as fluffy white spots. The disease especially affects the leaves of Cucurbitaceae crops, generally at the end of the season. Not to be confused with "real" mildew, which affects potatoes, tomatoes, and other crops.

Pyrethrum: Insecticidal powder extracted from dried chrysanthemum flowers; slightly toxic to humans.

Ramial chipped wood (RCW): RCW is an uncomposted mix of chipped wood branches. By extension, the term also refers to a growing technique designed to restore humus-rich soil that imitates soil found in forests. The technique consists of introducing branches (up to 2¾ inches in diameter) of freshly cut and shredded wood.

Rotary harrow: A tool for shallow tillage, with teeth that turn on a vertical axis (unlike the Rototiller, whose teeth are on a horizontal axis). This piece of machinery is mainly used to prepare seedbeds.

Rotenone: An insecticide that is derived from the roots of plants in the Fabaceae family and has been used in agriculture for a long time. Its harmlessness is now being questioned.

Rototiller: A tilling machine that turns and mixes the soil by the action of bent teeth mounted on a horizontal axis.

Runoff: Drainage of any rainwater that does not soak into the soil or evaporate in the air. Runoff is one of the causes of soil erosion: the water carries away soil particles as it drains; the size of these particles depends on the flow rate and the slope.

Seeding: A group of plants grown from seed on a given date. Seedings can be started directly in the garden (direct seeding) or indoors to be transplanted at a later time (indoor seeding).

Seedling: A young plant with only a few leaves.

Shade cloth: Tarp or net with fairly tight mesh. Shade cloth is used to cover a crop to protect it from the sun and, by extension, from overheating. This material makes it possible to maintain the right conditions for cool-climate crops when temperatures are high.

Share: In community-supported agriculture (CSA), the "share" is part of the harvest distributed each week to farm members. A share is generally made up of 8 to 12 different vegetables and herbs.

Soil mix: The substrate for crops seeded indoors. It is made of soil mixed with minerals or decomposed plant or animal matter (e.g., peat moss, perlite, vermiculite, and compost).

Solanaceae: A botanical family that includes potatoes, tomatoes, peppers, eggplants, and ground cherries.

Striped cucumber beetle: The main insect pest affecting the Cucurbitaceae family. In the adult stage, the insect has a black head and a yellow or orange body with three black stripes.

Thinning: Removing certain plants to allow the remaining ones to develop better. The goal of this task (generally done by hand on one's knees!) is to achieve the optimal density for a direct-seeded crop. Also known as "pricking out."

Topping: Cutting off the main end of a plant so that it stops putting out leaves and instead concentrates its energy on fruiting. This technique allows fruit crops (tomato, pepper, cucumber, Brussels sprouts, etc.) to reach maturity sooner than expected.

Tops: The name for the leaves of certain vegetables, most notably carrots.

Transplanting: A horticultural technique in which plants are started indoors in soil mix and later planted in gardens.

Vegetable grower: A farmer who grows vegetables commercially, whether in greenhouses or in fields.

Vetch: Plant in the Fabaceae family grown as a green manure. Common vetch is an annual species, while hairy vetch can be grown as a biennial. Not to be confused with tufted vetch, which is a weed.

Water budget: The relationship between amount of water that accumulates in the soil from precipitation and the amount that is lost by evapotranspiration, over a defined period. The water budget makes it possible to estimate the amount of water available in the soil to meet plant needs.

Yurt: Round tent traditionally used as a house by the nomad peoples of central Asia (e.g., Mongolia). Made out of canvas or, better yet, acrylic, a yurt makes an excellent temporary shelter in a temperate climate.

Index

Chinese cabbage. *See* Asian greens.
clay soil, 22, 23. 45, 83
climate, 19–20, 23, 24–25, 26, 87, 119
cold rooms, 127, 128, 131–132, 152
Coleman, Eliot, 2, 3, 41, 81, 102, 119
collinear hoe, 102–103
Colorado potato beetle, 158
commercial compost, 62–64
commercial soil mix, 83
common vetch, 71, 73, 74
community, 143–145
community-shared agriculture (CSA), 1, 9, 16
    advantages of, 11
    and shares, 134–137
    and space requirements, 21
compaction, of soil, 23, 29, 41, 43, 47
compost, 7, 45, 53, 56, 58, 59, 60, 61–64, 75
conventional farming, 30, 111
copper sprays, 115
corn, 181
cotyledon stage, 103, 107
cover crops, 54, 70–78. *See also* ground cover, pre-crop.
crop calendar, 137–138
crop planning, 8, 13, 21, 91, 92, , 133–141
Crop Planning for Organic Vegetable Growers, 13
crop requirements, 55–57
crop residues, 42, 48, 50, 51, 58, 72
crop rotation, 53–54, 55, 60, 64–70, 76–78
crop spacing, 7, 8, 43, 101
cucumber beetle, 111, 156, 176
cucumbers, 116, 124, 127, 129, 155–157
Cucurbitaceae, 56, 66, 92, 114. *See also by name.*
curly endive, 154–155
customer loyalty, and CSA, 11

D

deer, protection against, 35–36
design, of market gardens, 31–39, 112
direct seeding, 37, 51, 95–100, 120–121

direct selling, 11–13
disease prevention, 114–116
diseases, 60, 65, 82, 90, 111–117. *See also* blossom-end
    rot; fungal diseases.
dogs, on farms, 36
dolomitic limestone, 60
drainage, 23, 24–26, 42–43, 58
drainage tiles, 25–26
drip irrigation, 37

E

Earth Pond A to Z, 26
EarthWay Precision Garden Seeders, 96, 149
earthworms, 42, 47, 50
ecological pest control, 112
eggplants , 157–158
equipment. *See by name.*
Equiterre, 1, 10, 11

F

fallow plots, 21, 66, 70, 73
fall rye, 72, 74, 108
false seedbeds, 50, 106. *See also* stale seedbed
    technique.
farmers' markets, 11, 18, 20, 134, 143
fencing, 36, 37, 152
fertilization, organic, 53–79
    and cover crops, 76–78
    dosages, 56, 57
    strategy, 55
field blocks, 34
field peas, 71
financial objectives, 133, 134
flail mowers, 42, 47–48
flame weeding, 106, 107–108, 153
flea beetles, 148, 151, 161, 162, 163, 168, 173, 181
floating row covers, 90, 92, 93, 95, 105, 120–122, 158, 163
foliar spray, 61
food distribution, conventional, 11, 144

seeding, for transplanting, 81–94
seeding density, 72, 73, 82, 99–100
seedling room, 85–86, 150
seedlings, 37, 150
seed stock, 95, 116, 152
shares, in CSA model, 134–137
single-harvest vegetables, 135. *See also by name.*
site selection, 17–30
site evaluation checklist, 18, 19
Six-Row Seeders, 97–99
slope, of land. *See* topography.
slugs, 108, 161, 165
snap peas, 173–174
snow peas, 173–174
soil amendments, 42, 43–44, 45. *See also* fertilization, organic.
soil blocking, 81
soil creation, 6–7, 45, 72
soil ecology, 78
soil fertility, management of, 57–61
soil inversion, 50, 102
soil mix, in cell flats, 82–84
soil preparation, 41–42, 50–52, 72, 75
soil quality, 21–22, 29–30, 45
soil sampling, 55
soil solarization, 106
soil testing, 22, 53, 55, 57 , 60
Solanaceae, 56, 66, 93, 114, 122. *See also by name.*
sorghum-Sudan grass hybrids, 72
space requirements, 20–21, 34–35, 44–45
spinach, 121, 174–175
spinosad, 113, 116
sprayers, 151–152
stale seedbeds, 65, 73, 105–106, 107, 138
start-up costs, 8–10
stirrup hoes, 102, 149
storage, 127–132
storage bins, 132
straw mulch, 108

succession crops, 7–8, 13, 70, 82, 135, 137, 148, 162, 176, 181
sulfur, 60
sulfur sprays, 115
summer onions, harvesting of, 129
summer squash, 127, 129, 135, 175–177
Swede midge, 151, 154, 163
Swiss chard, 161–162
symptoms, of plant diseases, 114–116
synthetic chemicals, 111
synthetic fertilizers, 29–30

T
tarnished plant bugs, 113, 140, 158, 167, 172
tarps, 42, 50, 63, 72, 73, 75, 104–105, 179
temperatures, for seedlings, 85, 86, 87–89
temporary structures, 28
Thériault, Fred, 13
thermal screening, 90
thermal weed control. *See* flame-weeding.
thermometers, 89
thinning, 96, 99
thrips, 114
tillage practices, 41–52, 73–74. *See also* broadforks.
tomatoes, 60, 86–87, 120, 124, 129, 177–180
topography, 22–24
trace elements. *See* micronutrients.
tractors, 44. *See also* two-wheel tractors.
transplanting, 81–94, 121
    advantages, 82
    table, 94
tunnel hoops, 121, 150
turnips, 180–181
two-wheel tractors, 45–47, 149–150

V
vacuum seeder, 86
Van Es, Harold, 58
vegetable oil, conversion kits, 153

About the Author

JEAN-MARTIN FORTIER and his wife MAUDE-HÉLÈNE DESROCHES are the founders of Les Jardins de la Grelinette, an internationally recognized micro-farm famous for its high productivity-profitability using low-tech, high-yield methods of production. A leading practitioner of biologically intensive cropping systems, Jean-Martin has more than a decade's worth of experience in mixed organic farming. He has written articles about his work for popular magazines such as *Canadian Organic Grower, La Terre de Chez Nous* and *Growing for Market*. He also contributes occasionally as a tool and equipment advisor for companies such as Johnny's Selected Seeds and Dubois Agrinovation. The original French language version of this book, *Le Jardinier-Maraîcher*, released in Fall 2012, has sold more than 15,000 copies.

If you have enjoyed *The Market Gardener*, you might also enjoy other

Books to Build a New Society

Our books provide positive solutions for people who
want to make a difference. We specialize in:

Sustainable Living ◆ **Green Building** ◆ **Peak Oil**
Renewable Energy ◆ **Environment & Economy**
Natural Building & Appropriate Technology
Progressive Leadership ◆ **Resistance and Community**
Educational & Parenting Resources

New Society Publishers
ENVIRONMENTAL BENEFITS STATEMENT

New Society Publishers has chosen to produce this book on recycled paper made
with 100% post consumer waste, processed chlorine free, and old growth free.

For every 5,000 books printed, New Society saves the following resources:[1]

31	Trees
2,837	Pounds of Solid Waste
3,121	Gallons of Water
4,071	Kilowatt Hours of Electricity
5,156	Pounds of Greenhouse Gases
22	Pounds of HAPs, VOCs, and AOX Combined
8	Cubic Yards of Landfill Space

[1] Environmental benefits are calculated based on research done by the Environmental Defense Fund and
other members of the Paper Task Force who study the environmental impacts of the paper industry.

For a full list of NSP's titles, please call 1-800-567-6772 or check out our web site at:

www.newsociety.com